Deflectometry and Image Denoising

Deflectometry and Image Denoising

Von der
Carl-Friedrich-Gauß-Fakultät
der Technischen Universität Carolo-Wilhelmina zu Braunschweig

zur Erlangung des Grades einer
Doktorin der Naturwissenschaften (Dr. rer. nat.)

genehmigte Dissertation

von
Birgit Komander
geboren am 24.12.1987
in Hannover

Eingereicht am: 19.12.2018
Disputation am: 13.03.2019
1. Referent: Prof. Dr. Dirk Lorenz
2. Referent: Prof. Dr. Michael Möller

2019

Bibliografische Information der Deutschen Nationalbibliothek
Die Deutsche Nationalbibliothek verzeichnet diese Publikation in der
Deutschen Nationalbibliografie; detaillierte bibliografische Daten
sind im Internet über http://dnb.d-nb.de abrufbar.
1. Aufl. - Göttingen: Cuvillier, 2019
Zugl.: (TU) Braunschweig, Univ., Diss., 2019

© CUVILLIER VERLAG, Göttingen 2019
Nonnenstieg 8, 37075 Göttingen
Telefon: 0551-54724-0
Telefax: 0551-54724-21
www.cuvillier.de

ISBN 978-3-7369-9997-8
eISBN 978-3-7369-8997-9

Gewidmet Joachim Komander

DANKSAGUNG

Diese Arbeit entstand während meiner Tätigkeit als wissenschaftliche Mitarbeiterin am Institut für Analysis und Algebra der Technischen Universität Braunschweig.

Ein ganz besonderer Dank gilt meinem Doktorvater, Professor Dr. Dirk Lorenz, der mir die Möglichkeit gegeben hat, viele neue Erfahrungen zu sammeln und der mich zuvor in meinem Studium und danach auf dem Weg zu meiner Promotion hervorragend betreut hat. Danke für das unermüdliche Beantworten meiner Fragen, die fachlichen Diskussionen und die immer gute Zusammenarbeit, die mir immer viel Spaß gemacht hat.

Prof. Dr. Michael Möller von der Universität Siegen danke ich herzlich für die Übernahme des Korreferats und die immer netten Treffen auf Konferenzen.

Ein weiterer Dank geht an Marcus Petz und Marc Fischer, Mitarbeiter des Instituts für Produktionsmesstechnik (IPROM) der Technischen Universität Braunschweig. Durch die von ihnen erhaltenen Daten und Problemstellung, wurde der Grundstein für diese Arbeit gelegt.

Ein weiterer Faktor für das Gelingen der Arbeit war das hervorragende Arbeitsklima in der Arbeitsgruppe und dem Institut. Die Kaffeerunden haben stets für eine kurze auffrischende Verschnaufpause gesorgt, insbesondere die hilfreichen und des Öfteren auch unbekümmerten und scherzhaften Gespräche.

Zusätzlicher Dank gilt den eifrigen Korrekturlesern Fiona Gottschalk, Christoph Helmer, Hinrich Mahler, Marko Stautz und Morten Wesche, die die Anzahl der vorhandenen Fehler auf ein Minimum beschränkt haben.

Ein großes Danke geht an Fiona Gottschalk für ihre Freundschaft und ihren Beistand, insbesondere in der letzten Phase des Zusammenschreibens dieser Arbeit.

Besonders bedanken möchte ich mich bei meinen Eltern, die mich in jeder Lebenslage unterstützt haben und mich stets darin bestärkt haben das zu machen, was mir Freude bereitet.

Schließlich bedanke ich mich bei allen, die ich hier im Eifer des Gefechts vergessen habe namentlich zu erwähnen, die aber nicht vergessen sind, entschuldigt bitte.

Braunschweig, 03. April 2019
Birgit Komander

TABLE OF CONTENTS

List of Figures

LIST OF TABLES

CHAPTER 1

INTRODUCTION

Consider manufactured parts, such as screws, car doors, lenses, or mirrors for lasers, for example. All these manufactured parts have to go through quality inspections checking if there are unwanted bumps or scratches that should not be there. There are different methods to measure the manufactured parts. One that we will consider in this thesis is a deflectometric measurement process that deals with the measurement of specular objects. The output of such measurement processes is given in some raw data depending on the process. The goal is to describe the measured object exactly by the data. This is one example for a so-called inverse problem.

Another example that we want to consider are images. These images can be photographies or MRI scans, for example. A photography can be corrupted by noise. For example these unwanted signals can occur in images where the photo was taken in a too dark environment without a flash light. That can look like the image in Figure 1.1.

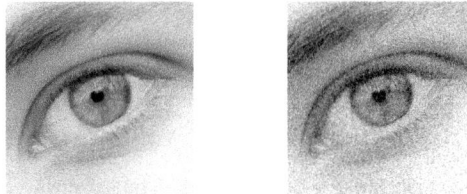

(a) Noise-free image.   (b) Image corrupted by noise.

Figure 1.1: Comparison of a noise-free image and an image corrupted by noise.

In general, in inverse problems an operator equation modeling a specific process is given. These processes are physical processes and applications include e.g. tomography, medical imaging, or object measurements. The aim is to find an input argument that provides the given results. To this end, an inversion of the operator equation is desired. However, these operators are usually not invertible. Additionally, inverse problems are highly sensitive to errors in the measurement data.

To overcome this issue, mathematical tools are needed to approximate an inversion of the problem. One approach is to construct functionals for which the minimization problem is well-posed in the sense that unique minimizers exist and are close to the unknown solution within a tolerance range. These functionals consist of two parts. The first part is a fidelity term which controls the deviation between the given output data and the data produced by the model for some input data. The second term is a regularization that gives the option to force the input parameters to fulfill certain properties such as how much an image resembles a "natural image".

The application that we consider in the first part of the thesis is a data fusion process. The given dataset is a result of a deflectometric measurement process [Pet04, Pet06] and is provided by the Institute of Production Measurement Technology (IPROM) at the Technical University Braunschweig.

Deflectometric measurement processes deal with object measurements of specular objects, such as lenses or mirrors. The aim is to calculate a dataset that describes the measured object exactly. The output data consists of two sets of separately measured types of data. There are the measured surface points and the measured surface orientation given by three spatial coordinates and by normal vectors, respectively. Since the direct measurement of the points is more sensitive to noise than the measurement of the normal vectors, the accuracy of data is inconsistent. In detail, the accuracy of the normal vectors is three orders of magnitude higher than the accuracy of the surface points. We resolve this issue with a data fusion process by solving a minimization problem which uses the normal vectors as a reference value. By doing so the accuracy of the surface points is increased.

Taking the gained insights we are able to develop new theories for image denoising. Image denoising, as we realized, is a problem similar to the data fusion process.

In imaging there are different methods to denoise an image. In 1992, Rudin, Osher and Fatemi introduced the total variation as a regularizer [ROF92],

$$\min_u \frac{1}{2} \int_\Omega |u - u_0|^2 \, \mathrm{d}x + \lambda \int_\Omega |\nabla u| \, \mathrm{d}x.$$

One problem in the resulting denoised images is the occurring staircasing effect, i.e. the creation of flat areas separated by jumps. One way to overcome this staircasing was proposed by Lysaker et al. in 2004 [LOT04]. The technique they proposed was a denoising of the image in two separate steps. In a first step, a total variation filter was used to smooth the normal vectors of the level sets of a given noisy image and then, as a second step, a surface was fitted to the resulting normal vectors. The method was designed in a dynamic way, i.e. by solving a certain partial differential equation to steady state.

A similar approach is taken in data fusion process. The measurement device does not only produce approximate point coordinates but also approximate surface normals. It turned out that the incorporation of the surface normals results in an effective, but fairly complicated and non-linear problem. In our approach we switch from surface normals to image gradients which leads to an effective method.

For image denoising we follow the idea of introducing additional information, e.g. gradient information, into the above ROF-model.

We formulate certain minimization problems in which use suitable reference values. In image denoising the reference value we want to use is an approximation of the image gradient vectors. Consequently, our approaches calculate such an approximation and use it as a reference value. Hence, our approaches are two-stage methods.

Another approach to prevent the staircasing effect is to go to higher orders of differentiation within the regularization term. One approach was proposed in 2010 namely the total generalized variation (TGV) functional [BKP10, KBPS11].

We propose different kinds of combinations of these functionals, since we can use the functionals as constraints or penalties. In this way we are able to formulate different minimization problems that are in some sense equivalent to the TGV problem. One advantage of some of these problems lies in the easy parameter choice rules that perform equally well as the TGV problem. Additionally, the duality gaps of these new problems are finite instead of infinite as it is usually the case in the primal-dual gap for the TGV problem. Hence, these can be used to create a reasonable stopping criterion for the optimization process. An additional advantage is the decreased runtime of the two-stage methods, since the problem is divided into two smaller problems.

1.1 Organization of the thesis

Chapter 2 provides underlying theory and notation that is used throughout this thesis. These preliminaries include notions of functional analysis such as functional spaces, e.g., spaces of bounded total variation and bounded total generalized variation, norms and seminorms, and different types of derivatives. Furthermore, the chapter contains some convex analysis, including illustrations of terms that are necessary for the solution theory of convex optimization problems. It also includes a brief overview of inverse problems in general and the direct method of the calculus of variations. The chapter concludes with solution methods for convex optimization problems that are based on [CNCP10]. Chambolle-Pock's primal-dual algorithm for solving minimization problems of the type

$$\min_{x \in X} F(x) + G(Kx)$$

for functionals F, G (e.g. representing the fidelity and the regularization term) with certain properties and a linear and continuous operator K closes this chapter.

The first of the two main parts of this thesis is Chapter 3. This chapter describes an inverse problem for an application in deflectometric measurements. The given data are sets of measured point coordinates and measured normal vectors of an object. These are provided by the same measurement setup but are determined independently. Because of the architecture of the measurement process and the different sensitivity to noise, the types of measured data do not have the same accuracy. In the chapter the geometry of the measurement setup is explained alongside the structure of the measured dataset. Moreover, different approaches to increase the accuracy to a higher order of magnitude which we call data fusion are discussed. Algorithms are proposed and tested on real datasets provided by the Institute of Production Measurement Technology (IPROM) at the Technical University Braunschweig.

The other part of the thesis, Chapters 4 to 6, uses the insights of the previous chapter and applies these to mathematical image denoising.

In Chapter 4 an idea of a two-stage image denoising method is proposed which is inspired by the fact that measured surface orientation is a powerful tool to lift the accuracy of measured surface point coordinates in the data fusion process. Consequently, a two-stage image denoising method is proposed which, in a first step, denoises the image gradients and takes these gradients into a second step as prior information where the image is denoised with respect to the solution from the gradient denoising step. Here, we propose two methods building on the same idea. Within each step the minimization functionals can be formulated via penalization or with constraints. Later on, the advantages and disadvantages of using one or the other formulation will be discussed.

Chapter 5 gives variants of total generalized variation image denoising. Taking the two-stage methods of Chapter 4, both steps can be combined into one optimization problem in various ways. The particular functionals considered in the two-stage methods can be combined into one optimization problem by pure penalization within the minimization functional, resulting in the total generalized variation problem [BKP10, KBPS11], or they can be combined in a mixed type using one or two of the particular functionals as constraints. In Chapter 5 the resulting combined methods are discussed. We investigate advantages and disadvantages of the problem formulation especially with respect to parameter choices.

In Chapter 6, the proposed methods for image denoising are experimentally tested on various images and different noise levels. Not all of the proposed combined methods come with an simple or clear parameter choice rule. The numerical experiments are restricted to those methods that do. Thus, first all methods are evaluated separately according to performance and quality and after that the methods are compared with one another.

PRELIMINARIES

This chapter introduces the mathematical background needed for this work and fixes the notation. We give a brief overview over Banach and dual spaces with examples and look into function spaces appropriate for solving minimization problems. Afterwards, we will recall convex analysis, inverse problems with variational calculus, and conclude with solving methods for convex optimization problems.

2.1 Functional analysis

In this section we will collect basic results and concepts. For more details and proofs the reader is referred to standard literature such as [Rud91, Bre].

In the following we denote by X a vector space over a field $\mathbb{K}$ ($\mathbb{R}$ or $\mathbb{C}$). A mapping $\|\cdot\|_X : X \to [0, \infty)$ is called *norm*, if the following holds:

1. $\|\lambda x\|_X = |\lambda| \, \|x\|_X \; \forall \lambda \in \mathbb{K}, \; \forall x \in X$,
2. $\|x + y\|_X \leq \|x\|_X + \|y\|_X \; \forall x, y \in X$,
3. $\|x\|_X = 0 \Rightarrow x = 0$.

If only 1. and 2. holds, the mapping is called a *seminorm*. Since we will work with norms coming from different vector spaces, lets review a few in the following example:

Example 2.1

1. Let $X = \mathbb{R}^d$. The following mappings define norms on $\mathbb{R}^d$:

$$\|x\|_p = \left(\sum_{k=1}^d |x_k|^p \right)^{1/p}, \quad 1 \leq p < \infty,$$

$$\|x\|_\infty = \max_{k=1,\dots,d} |x_k|.$$

2. Let $X = \ell^p$, i.e. the space of real-valued sequences for which the norm mappings are finite:

$$\|x\|_p = \left(\sum_{k=1}^\infty |x_k|^p \right)^{1/p}, \quad 1 \leq p < \infty,$$

$$\|x\|_\infty = \sup_{k=1,\dots,\infty} |x_k|.$$

Two norms $\|\cdot\|, \|\cdot\|'$ are called equivalent on X, if there are constants $c_1, c_2 > 0$ such that

$$c_1 \|x\| \leq \|x\|' \leq c_2 \|x\|$$

for all $x \in X$. If X is finite-dimensional, all norms on X are equivalent.

A normed (real) vector space $(X, \|\cdot\|_X)$ is called (real) *Banach space*, if it is *complete*. That means that every Cauchy sequence $(u_n)_{n\in\mathbb{N}}$ converges, i.e. there exists an $u \in X$ such that $\lim_{n\to\infty} \|u_n - u\|_X = 0$. We will write "normed vector space X" instead of the tuple above.

Definition 2.2 (Operator, Functional)
Let X, Y be normed vector spaces. A continuous linear mapping $K : X \to Y$ is called *operator*. If $Y = \mathbb{K}$, the mapping is called *functional*.

Definition 2.3 (Space of linear mappings)
Let X and Y be normed spaces. The *space of continuous linear mappings* is

$$\mathcal{L}(X,Y) := \{K : X \to Y \mid K \text{ is a linear and continuous operator}\}.$$

The operator is bounded by

$$\|K\|_{X\to Y} := \sup_{\|x\|_X \leq 1} \|Kx\|_Y < \infty. \tag{2.1}$$

$\mathcal{L}(X,Y)$ is a normed space with *operator norm* $\|\cdot\|_{X\to Y}$.

Theorem 2.4
Let Y be a Banach space. Then $\mathcal{L}(X,Y)$ is a Banach space, independently of the completeness of X.

Definition 2.5 (Lebesgue space $L^p(\Omega)$)
Let ν be the Lebesgue measure in $\mathbb{R}^d$, and $\Omega \subseteq \mathbb{R}^d$ open and nonempty. Then for $1 \leq p \leq \infty$

$$L^p(\Omega) := \left\{ u : \Omega \to \mathbb{C}, \nu\text{-measurable} \,\middle|\, \|u\|_{L^p(\Omega)} < \infty \right\}$$

equipped with the norm

$$\|u\|_{L^p(\Omega)} := \begin{cases} \left(\int_\Omega |u(x)|^p \, d\nu(x) \right)^{1/p}, & p \in [1,\infty), \\ \operatorname*{ess\,sup}_{x\in\Omega} |u(x)| := \inf \{\alpha \geq 0 \mid \nu(\{|u| > \alpha\}) = 0\}, & p = \infty \end{cases}$$

is a Banach space. Functions $u \in L^p(\Omega)$ are called p-integrable functions and $u \in L^\infty(\Omega)$ are essentially bounded measurable functions.

Per se the Lebesgue spaces (actually written as $\mathcal{L}^p(\Omega)$) are not normed vector spaces, since it is only equipped with a seminorm $\|\cdot\|^*_{\mathcal{L}^p(\Omega)}$. But by considering $N_p = \{u \mid u = 0 \ \nu\text{-a.e.}\}$, the kernel of $\|\cdot\|^*_{\mathcal{L}^p(\Omega)}$, and identifying functions which are equal ν-a.e., i.e. considering equivalence classes $[u]$ instead of u, we obtain

$$L^p(\Omega) = \mathcal{L}^p(\Omega) \big/ N_p.$$

These are the Banach spaces in Definition 2.5. We write u instead of $[u]$, and when there are no misunderstandings within the context, we will write $\|\cdot\|_p$ instead of $\|\cdot\|_{L^p(\Omega)}$, for $1 \leq p \leq \infty$. If $K : X \to Y$ is linear, then continuity of K is equivalent to

$$\|Kx\|_Y \leq c\|x\|_X$$

for all $x \in X$ and a constant $c > 0$. Due to this, a linear continuous operator is also called a *bounded* linear operator. The operator norm $\|K\|_{X \to Y}$ defined in Equation (2.1) is the smallest possible constant c to satisfy this inequality.

Definition 2.6 (Dual space, duality pairing)
Let X be a Banach space. The space $X^* := \mathcal{L}(X, \mathbb{K})$ of linear continuous functionals on a normed space X is called *dual space* of X. It is equipped with the dual norm

$$\|x^*\|_{X^*} := \sup_{\|x\|_X = 1} |x^*(x)| = \sup_{\|x\|_X \leq 1} |x^*(x)| = \sup_{x \in X \setminus \{0\}} \frac{|x^*(x)|}{\|x\|_X} \tag{2.2}$$

where

$$\langle x^*, x \rangle_{X^* \times X} = x^*(x). \tag{2.3}$$

The functional (2.3) is called the *duality pairing*. X^* also is a Banach space (which is a direct consequence of Theorem 2.4).

This definition immediately implies

$$\langle x^*, x \rangle_{X^* \times X} \leq \|x^*\|_{X^*} \|x\|_X \quad \text{for all } x \in X, \, x^* \in X^*. \tag{2.4}$$

Definition 2.7 (Adjoint Operator)
Let X and Y be Banach spaces, and $K \in \mathcal{L}(X, Y)$. The *adjoint operator* $K^* \in \mathcal{L}(Y^*, X^*)$ is defined by the relation

$$\langle K^*y, x \rangle_X = \langle y, Kx \rangle_Y \tag{2.5}$$

for all $x \in X$ and $y \in Y^*$. Further, it holds that $\|K^*\|_{Y^* \to X^*} = \|K\|_{X \to Y}$.

Example 2.8 (Dual spaces, dual pairs)

1. Let $X = \ell^p$, $1 < p < \infty$, then the dual space can be identified with $X^* \cong \ell^q$, where $\frac{1}{p} + \frac{1}{q} = 1$. The duality pairing is given by

$$\langle x^*, x \rangle_{X^* \times X} = \sum_{k=1}^{\infty} x_k^* x_k.$$

For $p = 1$ it is $(\ell^1)^* = \ell^\infty$, but $(\ell^\infty)^*$ is not a sequence space.

2. Let $X = L^p(\Omega)$, $1 < p < \infty$, then $(L^p(\Omega))^* \cong L^q(\Omega)$ for $\frac{1}{p} + \frac{1}{q} = 1$. The duality pairing is given by

$$\langle u^*, u \rangle_{L^q(\Omega) \times L^p(\Omega)} = \int_\Omega u^*(x)u(x)\, dx.$$

Similar to the sequence space cases above, $(L^1(\Omega))^* \cong L^\infty(\Omega)$, but $(L^\infty(\Omega))^*$ cannot be identified with $L^1(\Omega)$.

Definition 2.9 (Topological terms)
Let X be a normed vector space. $U \subset X$ is called

1. *closed*, if for every convergent sequence $(x_n)_{n \in \mathbb{N}} \subset U$ it holds that $\lim_{n \to \infty} x_n = x \in U$,

2. *compact*, if every sequence $(x_n)_{n \in \mathbb{N}} \subset U$ contains a convergent subsequence $(x_{n_k})_{k \in \mathbb{N}}$ with $\lim_{k \to \infty} x_{n_k} = x \in U$.

Further, the open ball on X is defined as $B_{X,r}(x) := \{y \in X \mid \|x - y\|_X < r\}$ and the closed ball is $\bar{B}_{X,r}(x) := \{y \in X \mid \|x - y\|_X \leq r\}$. (We will drop the "$X$" in the subscript, if the according space is clear, i.e. we write $B_r(x)$.) With this $U \subset X$ is called

3. *open*, if for all $x \in U$ there exists an $r > 0$ with $B_r(x) \subset U$. Therefore, all $x \in U$ are interior points of U and $U = U^\circ$, where U° are all interior points of U,

4. *bounded*, if U is contained in $\bar{B}_r(x)$ for an $r > 0$,

5. *convex*, if for all $x, y \in U$ and $\lambda \in [0, 1]$ it holds that $\lambda x + (1 - \lambda)y \in U$.

The definition of the norm directly gives us the property that open and closed balls are convex sets. In normed vector spaces it also holds that the complement of a closed set is open and vice versa.

Definition 2.10 (Weak(-*) convergence)
Let X be a normed vector space. A sequence $(x_n)_{n \in \mathbb{N}} \subset X$ is said to *converge weakly* to $x \in X$, if

$$\lim_{n \to \infty} \langle x^*, x_n \rangle_{X^* \times X} = \langle x^*, x \rangle_{X^* \times X}$$

for all $x^* \in X^*$; we write $x_n \rightharpoonup x$.
A sequence $(x_n^*)_{n \in \mathbb{N}} \subset X^*$ *converges weak-** to a $x^* \in X^*$, if

$$\lim_{n \to \infty} \langle x_n^*, x \rangle_{X^* \times X} = \langle x^*, x \rangle_{X^* \times X}$$

for all $x \in X$; we write $x_n^* \overset{*}{\rightharpoonup} x^*$.

Note, that the convergence $x_n \to x$ is also called *strong convergence* in X. Further, it holds that if $x_n \to x$ and $x_n^* \overset{*}{\rightharpoonup} x^*$ or $x_n \rightharpoonup x$ and $x_n^* \to x^*$, then $\langle x_n^*, x_n \rangle_{X^* \times X} \to \langle x^*, x \rangle_{X^* \times X}$. The duality pairing, however, does not converge in general, i.e. not for $x_n^* \overset{*}{\rightharpoonup} x^*$ and $x_n \rightharpoonup x$. Terms like continuity, closedness of mappings and topological terms like closedness of sets and compactness in case of strong convergence transfer directly to weak(-*) continuity, etc.

Definition 2.11 (Separable, reflexive)
A normed vector space X is called

1. *separable*, if it contains a countable dense subset,

2. *reflexive*, if the canonical linear isometric mapping $i : X \to X^{**}$, $(i(x))(x^*) = x^*(x)$ is surjective.

By the Weierstraß approximation theorem, the spaces $L^p(\Omega)$, for $1 < p < \infty$, and $C(\bar{\Omega})$ are separable. For reflexive spaces it holds that $X \cong X^{**}$, but this property is not sufficient. The spaces ℓ^p and L^p for $1 < p < \infty$ are reflexive, but ℓ^1 is not.

Theorem 2.12 (Eberlein-Šmulyan)
Let X be a normed vector space. Then X is reflexive if and only if $\bar{B}_1(0)$ is weakly compact.

Theorem 2.13 (Banach-Alaoglu)
If X is a separable normed vector space, $\overline{B}_{X^*,1}(0)$ is weakly- compact.*

Definition 2.14 (Hilbert space)
Let X be a $\mathbb{K}$-vector space. A normed vector space X is called a *pre-Hilbert space*, if there is an inner product $\langle \cdot\,,\,\cdot \rangle_X$ defined on $X \times X$ with $\langle x,\,x \rangle_X^{1/2} = \|x\|_X$ for all $x \in X$. If $(X, \|\cdot\|_X)$ is complete, the space is called a *Hilbert space*.

We will drop the "X" if the context is clear. Inner products satisfy the *Cauchy-Schwarz inequality*:

$$|\langle x,\,y \rangle_X| \leq \|x\|_X \|y\|_X. \tag{2.6}$$

Remark 2.15. For $p = 2$ the space $L^2(\Omega)$ equipped with the inner product

$$\langle u,\,v \rangle_{L^2(\Omega)} = \int_\Omega u(x)\,\overline{v(x)}\,\mathrm{d}\nu(x)$$

is a Hilbert space.

Theorem 2.16 (Fréchet-Riesz)
Let X be a Hilbert space and $x^* \in X^*$. Then there exists a unique $y \in X$ with $x^*(x) = x_y^*(x) = \langle x^*,\,x \rangle_{X^* \times X} = \langle x,\,y \rangle_X$ for all $x \in X$. Further, $\|x^*\|_{X^*} = \|y\|_X$.

Definition 2.17 (Directional derivative)
Let $F : X \to Y$ be an operator or functional. The *directional derivative* at $x \in X$ in direction $h \in X$ is defined as

$$D_h F(x) := \lim_{t \searrow 0} \frac{F(x + th) - F(x)}{t} \in Y,$$

if the limit exists.

Definition 2.18 (Gâteaux derivative)
Let X, Y be normed vector spaces. A mapping $F : X \to Y$ is called *Gâteaux-differentiable at* $x \in X$, if $D_h F(x)$ exists for all $h \in X$ and

$$DF(x) : X \to Y,\ h \mapsto D_h F(x)$$

is a linear and bounded operator. $DF \in \mathcal{L}(X, Y)$ is its *Gâteaux-derivative*. Further, F is called *Gâteaux-differentiable*, if it is at every $x \in X$.

Definition 2.19 (Fréchet derivative)
Let X and Y be normed vector spaces. $F : X \to Y$ is called *Fréchet-differentiable at* $x \in X$, if there is $DF \in \mathcal{L}(X, Y)$ with

$$\lim_{\|h\|_X \to 0} \frac{\|F(x + h) - F(x) - DF(x)h\|_Y}{\|h\|_X} = 0.$$

The *(Fréchet-)derivative of F (at x)* is $F'(x) = DF(x)$ and if F is Fréchet-differentiable in every $x \in X$, F is called *Fréchet-differentiable*.

Note, that if $F : X \to \mathbb{R}$ then $F'(x) \in \mathcal{L}(X, \mathbb{R}) = X^*$, hence, for functionals the derivative is an element of the dual space. Another class of function vector spaces we will need are Sobolev spaces and the associated notions.

Definition 2.20 (Spaces of test functions)

Let $\Omega \subset \mathbb{R}^d$ be non-empty and open. The space $\mathcal{D}(\Omega)$ is called *space of test functions* and is defined by

$$\mathcal{D}(\Omega) = \{f \in C^\infty(\Omega) \,|\, \text{supp} f \subset \Omega \text{ compact}\}.$$

Definition 2.21 (Weak derivative)

Let $\Omega \subset \mathbb{R}^d$ be non-empty, open and connected, $u \in L^1_{\text{loc}}(\Omega)$, and $\alpha \in \mathbb{N}^d$ a multiindex. A function $v \in L^1_{\text{loc}}(\Omega)$ is called the α-th *weak derivative* of u if for every test function $\varphi \in \mathcal{D}(\Omega)$ it holds that

$$\int_\Omega v(x)\varphi(x)\,\mathrm{d}x = (-1)^{|\alpha|} \int_\Omega u(x)\partial^\alpha\varphi(x)\,\mathrm{d}x.$$

If all weak derivatives with $|\alpha| \leq m$ exist, u is said to be m times weakly differentiable.

Derivatives in the classic way are also weak derivatives.

Example 2.22

1. Let $\Omega = \mathbb{R}$ and $u(x) = |x|$. If we define the sign function as

$$\text{sign}(x) = \begin{cases} -1, & x < 0, \\ 0, & x = 0, \\ 1, & x > 0 \end{cases}$$

it is clear that $v(x) = \text{sign}(x)$ satisfies

$$\int_\Omega |x|\,\varphi'(x)\,\mathrm{d}x = -\int_\Omega \text{sign}(x)\varphi(x)\,\mathrm{d}x$$

with integration by parts.

2. Now, we want to differentiate the sign function, i.e. $u(x) = \text{sign}(x)$. For $x \neq 0$ the derivative is 0. But the constant zero function is not a good candidate as the derivative of sign, since the fundamental theorem of calculus is not satisfied for all $x \in \mathbb{R}$. For $a, b \in \mathbb{R}$ with $a < 0$ and $b > 0$ it is

$$\int_a^b \text{sign}'(x)\,\mathrm{d}x = 0 \neq 2 = \text{sign}(b) - \text{sign}(a).$$

However, we can calculate the following with integration by parts and $\varphi(x) \to 0$, $|x| \to \infty$:

$$\int_\mathbb{R} v(x)\varphi(x) = -\int_\mathbb{R} \text{sign}(x)\varphi'(x)\,\mathrm{d}x = 2\varphi(0) = 2\delta_0(x),$$

where $\delta_0 = \delta$ is the *delta distribution* in 0.

Definition 2.23 (Sobolev spaces)

For $1 \leq p \leq \infty$ and $m \in \mathbb{N}$

$$W^{m,p}(\Omega) = \{u \in L^p(\Omega) \,|\, D^\alpha \in L^p(\Omega), 0 \leq |\alpha| \leq m\}$$

is a *Sobolev space* of order m, p. Equipped with the norm

$$\|u\|_{W^{m,p}(\Omega)} = \begin{cases} \left(\sum_{|\alpha| \leq m} \|D^\alpha u\|^p_{L^p(\Omega)} \right)^{1/p}, & 1 \leq p < \infty, \\ \max_{|\alpha| \leq m} \|D^\alpha u\|_{L^\infty(\Omega)}, & p = \infty, \end{cases}$$

it is a Banach space. In case of $p = 2$ the space $H^m(\Omega) = W^{m,2}(\Omega)$ is a Hilbert space with inner product

$$\langle u, v \rangle_{H^m(\Omega)} = \sum_{|\alpha| \leq m} \langle D^\alpha u, D^\alpha v \rangle_{L^2(\Omega)}.$$

Definition 2.24
Let $(\Omega, \mathcal{A})$ be a measurable space. A function $\mu : \mathcal{A} \to \mathbb{R}^d$ is called *vector-valued measure* if

1. $\mu(\emptyset) = 0$,
2. μ is countable additive, i.e. for any sequence $(A_i)_{i \in \mathbb{N}}$ of disjoint sets in $\mathcal{A}$ such that the union is in $\mathcal{A}$, it holds that

$$\mu \left(\bigcup_{i=1}^\infty A_i \right) = \sum_{i=1}^\infty \mu(A_i).$$

Taking the supremum over all partitions of A the *variation* of a measure μ is defined by

$$|\mu|(A) = \sup \left\{ \sum_{i=1}^\infty |\mu(A_i)| \ \bigg| \ \bigcup_{i=1}^\infty A_i = A \right\}.$$

Definition 2.25 (Borel, Radon measure)
A vector-valued measure $\mu : \mathcal{A} \to \mathbb{R}^d$ is called *Borel measure* if $\mathcal{A}$ contains all Borel-sets. The set of all $\mathbb{R}^d$-valued Borel measures is denoted by $\mathfrak{M}(\Omega, \mathbb{R}^d)$. A measure $\mu : \mathcal{A} \to [0, \infty]$ is called *Radon-measure* if it is a Borel-measure and if

1. for all compact sets K: $\mu(K) < \infty$,
2. for all open sets V: $\mu(V) = \sup \{\mu(K) \,|\, K \subset V, K \text{ compact}\}$,
3. for all measurable A: $\mu(A) = \inf \{\mu(V) \,|\, A \subset V, V \text{ open}\}$.

It can be shown that $\mathfrak{M}(\Omega, \mathbb{R}^d)$ is a vector space. With $\|\mu\|_{\mathfrak{M}} = |\mu|(\Omega)$ it is even a Banach space.

Theorem 2.26 (Hölder's inequality)
Let $(\Omega, \mathcal{A}, \mu)$ be a measure space and let $p, q \in [0, \infty]$ with $\frac{1}{p} + \frac{1}{q} = 1$ and the convention "$\frac{1}{\infty} = 0$". Then, for all measurable functions f, g on Ω holds

$$\|fg\|_{L^1(\Omega)} \leq \|f\|_{L^p(\Omega)} \|g\|_{L^q(\Omega)}.$$

Theorem 2.27 (Egorov)
Let $(\Omega, \mathcal{A}, \mu)$ a measurable set with $\mu(\Omega) < \infty$, and let (f_n) be a sequence of measurable functions on Ω such that f_n converges point-wise to $f : \Omega \to \mathbb{R}$ almost everywhere. Then, for all $\varepsilon > 0$ there exists a set $M_\varepsilon \in \mathcal{A}$ with $\mu(\Omega \setminus M_\varepsilon) < \varepsilon$, such that (f_n) converges uniformly to f in M_ε.

Theorem 2.28 (Luzin)
Let $(\Omega, \mathcal{A}, \mu) = (\mathbb{R}^n, \mathcal{M}_n, \nu)$ be a measurable set with Lebesgue's σ-algebra $\mathcal{M}_n$ and Lebesgue measure ν, and let $f : \mathbb{R}^n \to \mathbb{R}$ be a measurable function with bounded support $\{f \neq 0\} = M$. Then, for all $\varepsilon > 0$ there exists a compact set $M_\varepsilon \subset M$ with $\nu(M \setminus M_\varepsilon) < \varepsilon$ such that the restriction $f \restriction_{M_\varepsilon}$ is continuous on M_ε.

2.2 Convex analysis

Since we want to work with minimization problems, the question of existence of minimizers is crucial. In order to answer this question, we will use variational calculus.

Let $\bar{\mathbb{R}} = \mathbb{R} \cup \{\infty\}$ and we define for all $a \in \mathbb{R}, a \leq \infty$:

$$a + \infty = \infty,$$
$$\infty \leq \infty,$$
$$\lambda > 0 \quad \Rightarrow \quad \lambda\infty = \infty.$$

Definition 2.29 (Domain, kernel, range, graph)
Let X, Y be normed vector spaces, $U \subset X$ a real subset of X and $F : U \to Y$ be a mapping. The effective *domain* of a functional $F : X \to \bar{\mathbb{R}}$ is defined as

$$\operatorname{dom} F := \{x \in X \mid F(u) < \infty\}. \tag{2.7}$$

The *kernel* of F is defined as

$$\ker F := \{x \in U \mid F(u) = 0\}. \tag{2.8}$$

The *range* of F is defined as

$$\operatorname{ran} F := \{F(x) \in Y \mid x \in U\} \tag{2.9}$$

and the *graph* as

$$\operatorname{graph} F := \{(x, y) \in X \times Y \mid y = F(x)\}. \tag{2.10}$$

Definition 2.30 (Set-valued mappings)
Let X and Y be normed spaces. A *set-valued mapping* $F : X \rightrightarrows Y$ is a subset of $X \times Y$. We write $F(x) = \{y \in Y \mid (x, y) \in F\}$ and will use $y \in F(x)$ instead of $(x, y) \in F$.

Definition 2.31 (Properties of functionals)
Let X be a Banach space. A functional $F : X \to \bar{\mathbb{R}}$ is called

1. *proper*, if $\operatorname{dom} F \neq \emptyset$,
2. *coercive*, if for $\|x_n\|_X \to \infty$ it follows that also $F(x_n) \to \infty$, as $n \to \infty$,
3. *convex*, if for all $x, y \in X$ and $\lambda \in [0, 1]$ it holds that

$$F(\lambda x + (1 - \lambda)y) \leq \lambda F(x) + (1 - \lambda)F(y), \tag{2.11}$$

4. *strictly convex* , if the above inequality is strict for $\lambda \in (0, 1)$,
5. *lower semi-continuous* (l.s.c.), if for all sequences $(x_n)_{n \in \mathbb{N}} \subset X$ and $x \in X$ with $x_n \to x$, $n \to \infty$ it holds that

$$F(x) \leq \liminf_{n \to \infty} F(x_n). \tag{2.12}$$

Example 2.32 (Absolute value function)
For $X = \mathbb{R}$ the function $F(u) = |u|$ is convex. Since by the triangle inequality it holds $|tu + (1 - t)v| \leq t|u| + (1 - t)|v|$.

Besides norm functions, another useful class of convex functions is given by the indicator function.

Definition 2.33 (Indicator function)
Let $C \subset X$ be a convex set. The function

$$\mathcal{I}_C(x) = \begin{cases} 0, & x \in C, \\ \infty, & x \notin C, \end{cases}$$

is called *indicator function of C*.

Using the terms of weak(-*) convergence, we define also weakly(-*) lower semi-continuous functionals via weakly(-*) convergent sequences.

Definition 2.34 (Epigraph)
Let X be a Banach space and $F : X \to \bar{\mathbb{R}}$. The *epigraph* of a functional is defined as

$$\operatorname{epi} F := \{(x, t) \in X \times \mathbb{R} \mid F(x) \leq t\}.$$

The properties defined in Definition 2.31 can also be characterized via the epigraph of the functional F.

Lemma 2.35
A functional $F : X \to \bar{\mathbb{R}}$ is

1. *proper, if and only if* $\operatorname{epi} F \neq \emptyset$,

2. *convex, if and only if* $\operatorname{epi} F$ *is convex,*

3. *lower semi-continuous, if and only if* $\operatorname{epi} F$ *is closed.*

Example 2.36
In Figure 2.1(a) the function is lower semi-continuous. The right limit lies above the left limit, hence, $\limsup_{x \to 0} F_1(x) \geq F_1(0)$. Would the upper point be the function value instead of 0, the function would not be lower semi-continuous. The epigraph of F_1 is not convex, hence, F_1 is not convex unlike F_2, where $\operatorname{epi} F_2$ is convex.

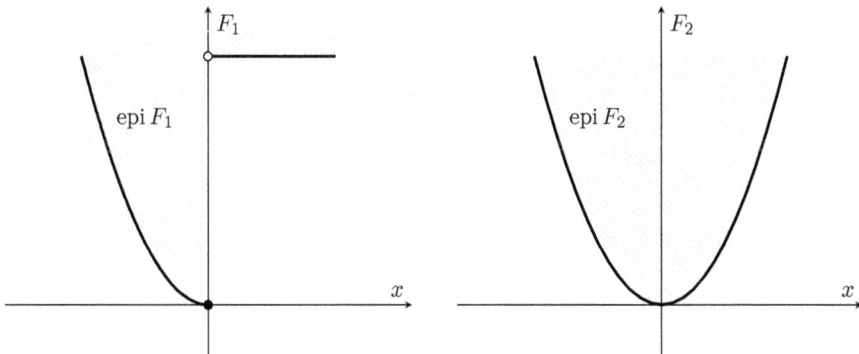

(a) Lower semi-continuous function. (b) Convex and lower semi-continuous function.

Figure 2.1: Example functions for lower semi-continuity and convexity by means of the epigraph of a function.

Lemma 2.37 (Lower semi-continuity under operations)
Let X and Y be Banach spaces and $F : X \to \bar{\mathbb{R}}$ be weakly lower semi-continuous. Then, the following functionals are also weakly lower semi-continuous:

1. *αF for all $\alpha \geq 0$,*
2. *$F + G$ for $G : X \to \bar{\mathbb{R}}$ weakly lower semi-continuous,*
3. *$\varphi \circ F$ for $\varphi : \bar{\mathbb{R}} \to \bar{\mathbb{R}}$ lower semi-continuous and strictly increasing,*
4. *$F \circ \psi$ for $\psi : Y \to X$ weakly continuous, i.e. for $y_n \rightharpoonup y$ it holds $\psi(y_n) \rightharpoonup \psi(y)$,*
5. *$x \mapsto \sup_{i \in I} F_i(x)$ with $F_i : X \to \bar{\mathbb{R}}$ weakly lower semi-continuous for any set I.*

Note, that Lemma 2.37.5. does not hold for continuous functions. Similar to the lower semi-continuity of functions under operations (cf. Lemma 2.37), we also have properties for convexity.

Lemma 2.38 (Convexity under operations)
Let X and Y be normed vector spaces and let $F : X \to \bar{\mathbb{R}}$ be convex. Then, the following functionals are convex as well:

1. *αF for all $\alpha \geq 0$,*
2. *$F + G$ for $G : X \to \bar{\mathbb{R}}$ convex; if one of F or G is strictly convex, so is $F + G$,*
3. *$\varphi \circ F$ for $\varphi : \bar{\mathbb{R}} \to \bar{\mathbb{R}}$ convex and increasing,*
4. *$F \circ K$ for $K : Y \to X$ linear,*
5. *$x \mapsto \sup_{i \in I} F_i(x)$ with $F_i : X \to \bar{\mathbb{R}}$ convex for any set I.*

Corollary 2.39
Let X be a Banach space. Then $\| \cdot \|_X$ is proper, coercive, and weakly lower semi-continuous.

Proof. From the definition follows directly the coercivity and $\operatorname{dom} \| \cdot \|_X = X$. Since

$$\|x\|_X = \sup_{\|x^*\|_{X^*} \leq 1} \left| \langle x^*, x \rangle_{X^* \times X} \right|$$

and the supremum is attained within a Banach space the statement follows with Lemma 2.37.5.
$\square$

Lemma 2.40
Let $f : \mathbb{R} \to \bar{\mathbb{R}}$ be proper, convex, and lower semi-continuous. If $\Omega \subset \mathbb{R}^d$ is bounded and $1 \leq p \leq \infty$ then $F : L^p(\Omega) \to \mathbb{R}$ with

$$F(u) = \begin{cases} \int_\Omega f(u(x)) \, \mathrm{d}x, & \text{if } f \circ u \in L^1(\Omega), \\ \infty, & \text{else,} \end{cases}$$

is also proper, convex, and lower semi-continuous.

Definition 2.41 (Subdifferential, subgradient)
Let X be a real normed vector space and $F : X \to \bar{\mathbb{R}}$ a convex functional. An element $z \in X^*$ is called *subgradient* if

$$F(x) + \langle z, y - x \rangle_X \leq F(y) \quad \forall y \in X.$$

The *subdifferential* $\partial F(x)$ of F in x is the set of all subgradients:

$$\partial F(x) = \{ z \mid F(x) + \langle z, y - x \rangle_X \leq F(y) \quad \forall y \in X \} \subset X^*.$$

Example 2.42 (Subdifferential in $\mathbb{R}$)

1. $F_1(x) = |x|$.

   The subgradient is defined as $\partial F_1(x) = \{s \in \mathbb{R} \mid F_1(x) + s(y - x) \le F_1(y) \, \forall y \in \mathbb{R}\}$. For $x > 0$ the inequality holds if $s = 1$. Similar, for $x < 0$ the subgradient is given with $s = -1$. For $x = 0$, the inequality reads $sy \le |y|$ and this hold for all $s \in [-1, 1]$. The subdifferential is (cf. Figure 2.2)

   $$\partial F_1(x) = \partial |x| = \begin{cases} \{-1\}, & x < 0, \\ [-1, 1], & x = 0, \\ \{1\}, & x > 0. \end{cases}$$

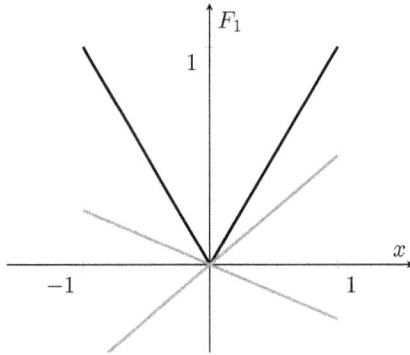

(a) Absolute value function. The gray lines are two of the affine supporting lines.

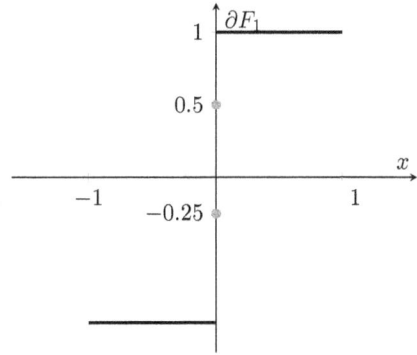

(b) The subdifferential of the absolute value function. The gray dots correspond to the affine supporting lines of $|x|$.

Figure 2.2: Subdifferential of the convex absolute value function on $[-1, 1]$ in $\mathbb{R}$.

2. $F_2(x) = \mathcal{I}_{[-1,1]}(x) = \begin{cases} 0, & x \in [-1, 1], \\ \infty, & \text{else.} \end{cases}$

   For $x \in (-1, 1)$ the function is constant zero, hence the derivative is zero as well. For $x \in [-\infty, -1) \cup (1, \infty]$ the derivative does not exist. Consider the point $x = 1$. There is $s(y - 1) \le 0$ for all $y \in [-1, 1]$ if $s \ge 0$, therefore $\partial F_2(1) = [0, \infty)$. Analogously, we get $\partial F_2(-1) = (-\infty, 0]$, and the subgradient of an indicator function is (cf. Figure 2.3)

   $$\partial F_2(x) = \partial \mathcal{I}_{[-1,1]}(x) = \begin{cases} (-\infty, 0], & x = -1, \\ \{0\}, & x \in (-1, 1), \\ [0, \infty), & x = 1, \\ \emptyset, & \text{else.} \end{cases}$$

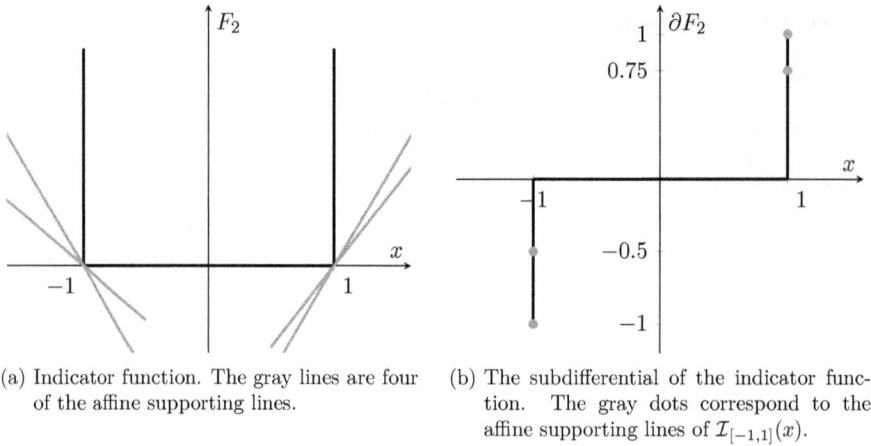

(a) Indicator function. The gray lines are four of the affine supporting lines.

(b) The subdifferential of the indicator function. The gray dots correspond to the affine supporting lines of $\mathcal{I}_{[-1,1]}(x)$.

Figure 2.3: Subdifferential of an indicator function on $[-1, 1]$ in $\mathbb{R}$.

Definition 2.43 (Convex conjugate)
Let X be a real topological space and let X^* be the dual space to X. For a functional $F : X \to \overline{\mathbb{R}}$ the *convex conjugate* $F^* : X^* \to \overline{\mathbb{R}}$ is defined by

$$F^*(x^*) = \sup_{x \in X} \langle x^* , x \rangle_{X^* \times X} - F(x).$$

Example 2.44 (Dual function of the absolute value function)
For $F : \mathbb{R} \to \mathbb{R}$, $F(x) = |x|$ the convex conjugate is

$$F^*(x^*) = \sup_{x \in \mathbb{R}} x^* x - |x|$$

and also $|x^*| |x| \geq x^* x$. For $|x^*| \leq 1$ it is

$$F^*(x^*) = \sup_{x \in \mathbb{R}} x^* x - |x| = 0,$$

since $x^* x \leq |x^*| |x| \leq |x|$ and $x^* x - |x|$ maximal for $x = 0$. For $|x^*| > 1$ set $t \geq 0$. Then

$$F^*(x^*) = \sup_{x \in \mathbb{R}} t\, x^* x - |tx| = \sup_{x \in \mathbb{R}} t(x^* x - |x|) = +\infty,$$

for $t \to \infty$, since $x^* x - |x| > 0$. The convex conjugate is $F^*(x^*) = \mathcal{I}_{|\cdot| \leq 1}(x^*) = \mathcal{I}_{[-1,1]}(x^*)$.

Lemma 2.45
Let $F : X \to \overline{\mathbb{R}}$ be convex and proper. Then

$$p \in \partial F(u) \quad \Leftrightarrow \quad F(u) + F^*(p) = \langle p, u \rangle_{X^* \times X} .$$

Proof. Definition 2.43 of the dual function implies

$$F(u) + F^*(p) \geq \langle p, u \rangle_{X^* \times X}$$

for all u, p. If $p \in \partial F(u)$, then $F(u) + \langle p, v - u \rangle_{X^* \times X} \leq F(v)$ for all $v \in X$. Therefore,

$$
\begin{aligned}
F(u) + \langle p, v - u \rangle_{X^* \times X} \leq F(v) \quad &\Leftrightarrow \quad \langle p, u \rangle_{X^* \times X} - F(u) \geq \langle p, v \rangle_{X^* \times X} - F(v) \\
&\Leftrightarrow \quad \langle p, u \rangle_{X^* \times X} - F(u) \geq \underbrace{\sup_{v \in X} \langle p, v \rangle_{X^* \times X} - F(v)}_{=F^*(p)} \\
&\Leftrightarrow \quad \langle p, u \rangle_{X^* \times X} \geq F(u) + F^*(p).
\end{aligned}
$$

$\square$

Theorem 2.46 (Inversion rule for subgradients)
Let X be a Banach space and $F : X \to \bar{\mathbb{R}}$ be proper, convex and lower semi-continuous. Then

$$(\partial F)^{-1} = \partial(F^*).$$

Proof. It is $u \in (\partial F)^{-1}(p)$ if and only if $p \in \partial F(u)$. Together with Lemma 2.45 we get

$$
\begin{aligned}
u \in (\partial F)^{-1}(p) \quad &\Leftrightarrow \quad p \in \partial F(u) \\
&\Leftrightarrow \quad F(u) + F^*(p) = \langle p, u \rangle_{X^* \times X} \\
&\Leftrightarrow \quad u \in \partial F^*(p).
\end{aligned}
$$

$\square$

Remark 2.47. In Example 2.42 we calculated the subdifferentials of the absolute value function and the indicator function on the set $[-1, 1]$. With Theorem 2.46 we are able to calculate these by invertion; it is $\partial(|u|) = \mathrm{sign}(u)$ and therefore, since $(|u|)^* = \mathcal{I}_{[-1,1]}(u)$ (cf. Example 2.32), it is $\partial((|u|)^*) = (\partial(|u|))^{-1} = (\mathrm{sign})^{-1}(u)$.

Example 2.48 (Dual functions of indicator functions)
Let $C \subset X$ be a convex set and $F : X \to \bar{\mathbb{R}}$, $F(x) = \mathcal{I}_C(x)$. The dual function is

$$
\begin{aligned}
F^*(x^*) &= \sup_{x \in X} \langle x^*, x \rangle_{X^* \times X} - \mathcal{I}_C(x) \\
&= \sup_{x \in C} \langle x^*, x \rangle_{X^* \times X}.
\end{aligned}
$$

Lemma 2.49
Let $F : X \to \bar{\mathbb{R}}$, $F(x) = \|x\|_X$. Then, the conjugate of F is given by

$$
F^*(x^*) = \mathcal{I}_{\|\cdot\|_{X^*} \leq 1}(x^*) = \begin{cases} 0, & \|x^*\|_{X^*} \leq 1, \\ \infty, & else. \end{cases}
$$

Proof. The dual norm of a norm $\|\cdot\|_X$ is defined by

$$\|x^*\|_{X^*} = \sup_{\|x\|_X \leq 1} \langle x, x^* \rangle.$$

We can write this as

$$
\begin{aligned}
\|x^*\|_{X^*} = \sup_{x \neq 0} \frac{\langle x, x^* \rangle}{\|x\|_X} &\geq \frac{\langle x, x^* \rangle}{\|x\|_X}, \quad x \neq 0 \\
&\Leftrightarrow \langle x, x^* \rangle \leq \|x\|_X \|x^*\|_{X^*}
\end{aligned}
$$

Further, the conjugate function is defined by

$$F^*(x^*) = \sup_{x \in X} \langle x, x^* \rangle - F(x).$$

We consider the case distinction:

1. Let $\|x^*\|_{X^*} \leq 1$. This yields

$$\langle x, x^* \rangle \leq \|x\| \, \|x^*\|_{X^*} = \|x\|_X, \ \forall x \in X.$$

Both sides are equal if $x = 0$. For the conjugate follows

$$\begin{aligned} F^*(x^*) &= \sup_{x \in X} \langle x, x^* \rangle - F(x) \\ &= \sup_{x \in X} \langle x, x^* \rangle - \|x\|_X \\ &= 0 \end{aligned}$$

since the term $\langle x, x^* \rangle - \|x\|_X$ is maximal at $\|x\|_X = 0$.

2. Let $\|x^*\|_{X^*} > 1$. Since $\|x^*\|_{X^*} = \sup_{\|x\|_X \leq 1} \langle x, x^* \rangle$, there exists an $x \in X$ with $\|x\|_X \leq 1$ such that $\langle x, x^* \rangle > 1$. For some $t \leq 0$ the computation of the conjugate writes as

$$\begin{aligned} F^*(x^*) &= \sup_{x \in X} \langle t\,x, x^* \rangle - \|t\,x\|_X \\ &\geq t \, \langle x, x^* \rangle - t \, \|x\|_X \\ &= t \, \underbrace{(\langle x, x^* \rangle - \|x\|_X)}_{>0} \end{aligned}$$

For $t \to \infty$ the last term will diverge to ∞ and it follows that in this case $F^*(x^*) = \infty$.

$\square$

Lemma 2.50 (Subdifferential of a norm)
Let X be a Banach space with given norm $\|\cdot\|_X$. The subdifferential then is

$$\partial(\|\cdot\|_X)(x) = \left\{ x^* \in X^* \,\middle|\, \langle x^*, x \rangle_{X^* \times X} = \|x\|_X, \ \|x^*\|_{X^*} \leq 1 \right\}.$$

Proof. Let $\mathcal{G}(x) := \left\{ x^* \in X^* \,\middle|\, \langle x^*, x \rangle_{X^* \times X} = \|x\|_X, \ \|x^*\|_{X^*} \leq 1 \right\}$. (We write $\langle \cdot, \cdot \rangle$ and drop $X^* \times X$ for now.) We show both inclusions:

1. Let $x^* \in \mathcal{G}(x)$ and $y \in X$ arbitrary. Then it is, since $\langle x^*, x \rangle = \|x\|_X$, and by using Hölder's inequality:

$$\begin{aligned} \langle x^*, y - x \rangle + \|x\|_X &= \langle x^*, y \rangle - \langle x^*, x \rangle_{X^* \times X} + \|x\|_X \\ &= \langle x^*, y \rangle \\ &\leq \|x^*\|_{X^*} \|y\|_X \\ &\leq \|y\|_X, \end{aligned}$$

because $\|x^*\|_{X^*} \leq 1$.

2. Let $x^* \in \partial(\|\cdot\|_X)(x)$. For all $y \in X$ it holds that

$$\begin{aligned} \langle x^*, y - x \rangle + \|x\|_X &\leq \|y\|_X \\ \Leftrightarrow \quad \langle x^*, y \rangle - \langle x^*, x \rangle + \|x\|_X &\leq \|y\|_X \\ \Leftrightarrow \quad \langle x^*, y \rangle - \|y\|_X &\leq \langle x^*, x \rangle - \|x\|_X. \end{aligned}$$

Since this holds for all y we take the supremum and get

$$\sup_{y \in X} \langle x^*, y \rangle - \|y\|_X \leq \langle x^*, x \rangle - \|x\|_X.$$

The left-hand side is the convex conjugate of the dual norm, i.e. the indicator function on $\{\|x^*\|_{X^*} \leq 1\}$ (cf. Lemma 2.49). This side is infinite if $\|x^*\|_{X^*} > 1$. But since the right-hand side is finite for a fixed $x \in X$, it holds $\|x^*\|_{X^*} \leq 1$. Hence, again with Hölder's inequality

$$0 \leq \langle x^*, x \rangle - \|x\|_X$$
$$\leq \|x^*\|_{X^*} \|x\|_X - \|x\|_X$$
$$= \underbrace{(\|x^*\|_{X^*} - 1)}_{\leq 0} \underbrace{\|x\|_X}_{\geq 0} \leq 0.$$

Thus, we have $\|x^*\|_{X^*} \leq 1$ and $\langle x^*, x \rangle = \|x\|_X$.

$\square$

2.3 Functional spaces - Spaces of bounded (generalized) variation

Let us review the total variation of a function and the space of bounded total variation.

Definition 2.51 (BV, total variation)
A function $u : \Omega \to \mathbb{R}$ is said to have *bounded total variation* if $u \in L^1(\Omega)$ and its distributional derivative satisfies $\nabla u \in \mathfrak{M}(\Omega, \mathbb{R}^d)$, i.e. ∇u is a Radon measure. The set of all functions with bounded total variation is denoted by $\mathrm{BV}(\Omega)$. On $\mathrm{BV}(\Omega)$ we define the *total variation* as

$$\mathrm{TV}(u) = \|\|\nabla u\|\|_{\mathfrak{M}} = \sup \left\{ \int_\Omega u \operatorname{div} \varphi \, \mathrm{d}x \,\middle|\, \varphi \in \mathcal{C}_0^1(\Omega, \mathbb{R}^d), |\varphi(x)| \leq 1 \, \forall x \in \Omega \right\}$$

and the norm on $\mathrm{BV}(\Omega)$ by

$$\|u\|_{\mathrm{BV}} = \|u\|_{L^1} + \mathrm{TV}(u).$$

If $u \in W^{1,1}(\Omega)$ then $\nabla u \in L^1(\Omega)$. Thus, it holds

$$\mathrm{TV}(u) = \sup \left\{ \int_\Omega \nabla u \varphi \, \mathrm{d}x \,\middle|\, \varphi \in \mathcal{C}_0^1(\Omega), |\varphi(x)| \leq 1 \, \forall x \in \Omega \right\}$$

With approximation of $\frac{\nabla u}{|\nabla u|}$ with a C^1-function φ with $|\varphi(x)| \leq 1$ we get

$$\mathrm{TV}(u) = \sup \left\{ \int_\Omega \nabla u \varphi \, \mathrm{d}x \,\middle|\, \varphi \in \mathcal{C}_0^1(\Omega), |\varphi(x)| \leq 1 \, \forall x \in \Omega \right\}$$
$$= \int_\Omega |\nabla u| \, \mathrm{d}x.$$

Indeed, functions with jump discontinuities are in the space BV. Thus, it seems reasonable to use this space in imaging applications. In 2010 K. Bredies et al. [BKP10] introduced a generalization to the total variation. This total generalized variation functional and the related function space have some advantages over the total variation.

Before we define the total generalized variation, we briefly introduce spaces of symmetric tensor fields (see [BG12, Chapter 2] and [BKP10, Section 2]).

Let $d \geq 1$ be the dimension, typically 2 or 3 in applications. Further, let

$$\mathcal{T}^k(\mathbb{R}^d) = \{\xi : \underbrace{\mathbb{R}^d \times \cdots \times \mathbb{R}^d}_{k \text{ times}} \to \mathbb{R} \mid \xi \ k\text{-linear}\}$$

for $k \in \mathbb{N}$. The k-linearity denotes linearity in every component. For $k = 0$, it is $\mathcal{T}^0(\mathbb{R}^d) = \mathbb{R}$.

Definition 2.52 (Symmetric k-tensors)

A tensor $\xi \in \mathcal{T}^k(\mathbb{R}^d)$ is called symmetric if $\xi(a_1, ..., a_k) = \xi(\pi(a_1), ..., \pi(a_k))$ for all $\pi \in S_k$, with S_k permutation group over $\{1, ..., k\}$. The vector space of symmetric k-tensors is defined as

$$\mathrm{Sym}^k(\mathbb{R}^d) = \{\xi : \underbrace{\mathbb{R}^d \times \cdots \times \mathbb{R}^d}_{k \text{ times}} \to \mathbb{R} \mid \xi \text{ } k\text{-linear and symmetric}\}.$$

For $k = 0$ it corresponds to scalar values. For $k = 1$ we have $\mathrm{Sym}^1(\mathbb{R}^d) = \mathbb{R}^d$, and for $k = 2$ we have the symmetric matrices, $\mathrm{Sym}^2(\mathbb{R}^d) = S^{d \times d}$.

Note the three basic operations for tensors: The tensor product. Let $\xi \in \mathcal{T}^k(\mathbb{R}^d)$, $\eta \in \mathcal{T}^l(\mathbb{R}^d)$, then

$$(\xi \otimes \eta)(a_1, ..., a_{k+l}) = \xi(a_1, ..., a_k)\eta(a_{k+1}, ..., a_{k+l}) \in \mathcal{T}^{k+l}(\mathbb{R}^d).$$

The trace $\mathrm{tr}(\xi) \in \mathcal{T}^{k-2}(\mathbb{R}^d)$ for $k \geq 2$, which is defined as

$$\mathrm{tr}(\xi)(a_1, ..., a_k) = \sum_{i=1}^{d} \xi(e_i, a_1, ..., a_{k-2}, e_i),$$

where e_i is the i-th canonical basic vector. Further, the symmetrization of any k-tensor $\xi \in \mathcal{T}^k(\mathbb{R}^d)$ is

$$(|||\xi)(a_1, ..., a_k) = \frac{1}{k!} \sum_{\pi \in S_K} \xi(a_{\pi(1)}, ..., a_{\pi(k)}).$$

The symmetrization of a tensor is a projection, i.e. $|||^2\xi = |||\xi$.

Consequently, the spaces $\mathcal{T}^k(\mathbb{R}^d)$ and $\mathrm{Sym}^k(\mathbb{R}^d)$ can be equipped with the inner product

$$\xi \cdot \eta = \mathrm{tr}^k(\bar{\xi} \otimes \eta)$$

with $\bar{\xi}(a_1, ..., a_k) = \xi(a_k, ..., a_1)$ and $|\xi| = \sqrt{\xi \cdot \xi}$.

Canonically, for $k = 0$ this leads to $|\xi| = \sqrt{\xi^2}$, since $\xi \in \mathbb{R}$, for $k = 1$ to the Euclidean norm, and for $k = 2$ we can identify $\xi \in \mathrm{Sym}^2(\mathbb{R}^d)$ with

$$\xi = \begin{pmatrix} \xi_{11} & \cdots & \xi_{1d} \\ \vdots & \ddots & \vdots \\ \xi_{d1} & \cdots & \xi_{dd} \end{pmatrix}, \quad |\xi| = \left(\sum_{i=1}^{d} \xi_{ii}^2 + 2\sum_{i<j} \xi_{i,j}^2 \right)^{1/2}.$$

Now, we are able to define Lebesgue spaces via symmetric k-tensor fields as mappings $\xi : \Omega \to \mathrm{Sym}^k(\mathbb{R}^d)$ where Ω is a fixed domain:

$$L^p(\Omega, \mathrm{Sym}^k(\mathbb{R}^d)) := \left\{ \xi : \Omega \to \mathrm{Sym}^k(\mathbb{R}^d) \text{ } \nu\text{-measurable} \mid \|\xi\|_{L^p(\Omega)} < \infty \right\}$$

with

$$\|\xi\|_{L^p(\Omega)} := \left(\int_\Omega |\xi|^p \, dx \right)^{1/p} \quad \text{for } 1 \leq p < \infty$$

and $\quad \|\xi\|_{L^\infty(\Omega)} := \underset{x \in \Omega}{\mathrm{ess \, sup}} \, |\xi(x)|.$

Since the vector norm in $\mathrm{Sym}^k(\mathbb{R}^d)$ is induced by an inner product, the usual duality relation for Lebesgue spaces holds, i.e. $(L^p(\Omega, \mathrm{Sym}^k(\mathbb{R}^d)))^* = L^q(\Omega, \mathrm{Sym}^k(\mathbb{R}^d))$ for $1 \leq p < \infty$ and $\frac{1}{p} + \frac{1}{q} = 1$.

We also extend the space of continuous functions to symmetric k-tensor fields:

$$C(\Omega, \mathrm{Sym}^k(\mathbb{R}^d)) := \left\{ \xi \in C(\overline{\Omega}), \xi \in \mathrm{Sym}^k(\mathbb{R}^d) \,\middle|\, \mathrm{supp}\xi \subsetneq \Omega \right\}.$$

We want to extend spaces of differentiable functions to symmetric k-tensor fields. This is more complicated since a differential tensor generally is not symmetric. But we can symmetrize these tensors. Let $D^l\xi : \Omega \to \mathcal{L}^l(\mathbb{R}^d, \mathrm{Sym}^k(\mathbb{R}^d))$ the l-th Fréchet-derivative of ξ and $\mathcal{L}^l(X, Y)$ being the space of l-linear and continuous mappings $X^l \to Y$. Then, we write a differentiable tensor as

$$\left(\nabla^l \otimes \xi\right)(x)(a_1, ..., a_{k+l}) = \left(D^l\xi(x)(a_1, ..., a_l)\right)(a_{l+1}, ..., a_{k+l}).$$

The symmetrization of the derivative denoted by $\mathcal{E}$, then is

$$\mathcal{E}^l(\xi) = |||(\nabla^l \otimes \xi) = (|||(\nabla \otimes))^l \xi.$$

Spaces of continuously differentiable functions in this sense are defined as

$$C^l(\overline{\Omega}, \mathrm{Sym}^k(\mathbb{R}^d)) = \left\{ \xi : \overline{\Omega} \to \mathrm{Sym}^k(\mathbb{R}^d) \,\middle|\, \nabla^m \otimes \xi \text{ continuous on } \overline{\Omega}, m = 0, ..., l \right\},$$

$$\|\xi\|_{l,\infty} = \max_{m=0,...,l} \|\mathcal{E}^m(\xi)\|_\infty.$$

Further, we define the l-divergence for elements of $C^l(\overline{\Omega}, \mathrm{Sym}^k(\mathbb{R}^d))$ as

$$\mathrm{div}^l(\xi) := \mathrm{tr}^l(\nabla^l \otimes \xi) \quad \text{with} \quad (\mathrm{div}^l \xi)_\beta = \sum_{\gamma \in \mathbb{N}^d, |\gamma|=l} \frac{l!}{\gamma!} \frac{\partial^l \xi_{\beta+\gamma}}{\partial x^\gamma}.$$

For $k = 0$, it is the identity, for $k = 1$ is the usual divergence. In case $k = 2$ it is with the interpretation as symmetric matrices.

$$(\mathrm{div}\,\xi)_i = \sum_{j=1}^d \frac{\partial \xi_{ij}}{\partial x_j}, \quad \mathrm{div}^2(\xi) = \sum_{i=1}^d \frac{\partial^2 \xi_{ii}}{\partial x_i^2} + 2\sum_{i<j} \frac{\partial^2 \xi_{ij}}{\partial x_i \partial x_j}.$$

This second order divergence will be used later on in the minimization functionals.

We now also can define Sobolev spaces with this notation. We get

$$W^{m,p}(\Omega, \mathrm{Sym}^l(\mathbb{R}^d)) := \left\{ \xi \in L^p(\Omega, \mathrm{Sym}^l(\mathbb{R}^d)) \,\middle|\, \mathcal{E}^k(\xi) \in L^p(\mathrm{Sym}^{l+k}(\mathbb{R}^d)), k = 0, ..., m \right\},$$

$$\|\xi\|_{W^{m,p}(\Omega, \mathrm{Sym}^l(\mathbb{R}^d))} = \left(\sum_{k=0}^m \|\mathcal{E}^k(\xi)\|_{L^p(\mathrm{Sym}^{l+k}(\mathbb{R}^d))}^p \right)^{1/p}, \quad 1 \le p < \infty,$$

and $\|\xi\|_{W^{m,p}(\Omega, \mathrm{Sym}^l(\mathbb{R}^d))} = \max_{k=0,...,m} \|\mathcal{E}^k(\xi)\|_{L^\infty(\mathrm{Sym}^{l+k}(\mathbb{R}^d))}.$

Definition 2.53 (Total generalized variation [BKP10])
Let $\Omega \subset \mathbb{R}^d$ be a domain, let $k \ge 1$, and let $\alpha_0, ..., \alpha_{k-1} > 0$. Then, the total generalized variation of order k with weight α for $u \in L^1_{\mathrm{loc}}(\Omega)$ is defined as the value of the functional

$$\mathrm{TGV}_\alpha^k(u) - \sup \left\{ \int_\Omega u \, \mathrm{div}^k v \, dx \,\middle|\, v \in C_{\mathcal{C}}^k(\Omega, \mathrm{Sym}^k(\mathbb{R}^d)), \|\mathrm{div}^l v\|_\infty \le \alpha_l, l - 0, ..., k-1 \right\},$$
(2.13)

where the supremum admits the value ∞ where the respective set is unbound from above.
The space

$$\mathrm{BGV}_\alpha^k(\Omega) = \left\{ u \in L^1(\Omega) \,\middle|\, \mathrm{TGV}_\alpha^k(u) < \infty \right\}, \quad \|u\|_{\mathrm{BGV}_\alpha^k} = \|u\|_1 + \mathrm{TGV}_\alpha^k(u), \quad (2.14)$$

is called the space of functions of bounded generalized variation of order k with weight α.

Remark 2.54. For $k = 1$ and $\alpha > 0$, we see that

$$\text{TGV}^1_\alpha(u) = \alpha \sup \left\{ \int_\Omega u \operatorname{div} v \, dx \ \Big| \ v \in C^1_C(\Omega, \text{Sym}^1(\mathbb{R}^d)), \|v\|_\infty \leq 1 \right\} = \alpha \, \text{TV}(u)$$

Let us summarize some properties of the TGV^k_α functional.

Lemma 2.55 (see [BKP10])
Let $k \in \mathbb{N}$ and $\alpha \in \mathbb{R}^k$, $\alpha > 0$. Then, the following statements hold:

1. *TGV^k_α is a seminorm on the normed space $\text{BGV}^k_\alpha(\Omega)$,*
2. *For $u \in L^1_{loc}(\Omega)$, $\text{TGV}^k_\alpha(u) = 0$ if and only if u is a polynomial of degree less than k,*
3. *TGV^k_α and $\text{TGV}^k_{\tilde\alpha}$ are equivalent for $\tilde\alpha \in \mathbb{R}^k$, $\tilde\alpha > 0$,*
4. *$\text{BGV}^k_\alpha(\Omega)$ is a Banach space,*
5. *TGV^k_α is proper, convex, and lower semi-continuous on each $L^p(\Omega)$, $1 \leq p \leq \infty$,*
6. *TGV^k_α is rationally invariant, i.e. for a rotation R it holds $\text{TGV}^k_\alpha(u) = \text{TGV}^k_\alpha(u \circ R)$,*
7. *$\text{TGV}^k_\alpha(u) \leq c \, \text{TV}(u)$ for any $u \in L^1_{loc}(\Omega)$ and some $c \in \mathbb{R}_{>0}$.*

2.4 Inverse problems

In general we distinguish between two different classes of problems. On the one hand we have the so-called *direct problem* where we calculate an effect out of a cause. For example we have given a differentiable function and we want to know the derivate. Then we just differentiate the function.

On the other hand we have the class of *inverse problems*. Here, we want to do the opposite. Given an effect, e.g. a point-cloud of a measured object with measurement uncertainties or a noisy image. The task is to recover the cause. For the given examples this would be the exact coordinates of the measured object or the noise-free image. Both problems will be considered later in this work.

Definition 2.56 (Well-posed, ill-posed problem [Had02])
A problem is said to be *well-posed* if

1. a solution exists (existence),
2. the solution is unique (uniqueness), and
3. the solution is continuous dependent on the initial data (stability).

If a problem is not well-posed, it is *ill-posed*.

In most of the problems we want to consider we have measured datasets, such as a point-cloud or a noisy image. Therefore the input data that is given and is in some way perturbed. These perturbations can be noise or measurement errors. In general, we will use u_0 for the given observed data from now on. The original unperturbed and unknown data is $u^\dagger$. In case of image denoising u_0 would be a noisy image and $u^\dagger$ the clean noise-free image. One model for measured datasets like an image writes as

$$u_0 = Ku + \eta \tag{2.15}$$

where K is some kind of image transformation such as blurring, sampling, or like a Radon transform for tomography and η is an additional white Gaussian noise. In the simplest situation we consider an average 0 and a standard deviation σ. Since η is a random variable according to the maximum likelihood principle we can find an approximation of $u^\dagger$ by solving the least-square minimization problem

$$\inf_u \mathcal{F}(u) = \inf_u \int_\Omega |Ku - u_0|^2 \, \mathrm{d}x \tag{2.16}$$

where Ω is the domain of the image and $\mathcal{F}$ is called the energy term.

The solution of the minimization problem may still be problematic and in fact it may still be ill-posed. In the next example we will see that the naive way of computing a solution u of a problem like $Ku = v$ can lead to poor reconstructions.

Example 2.57 (Differentiation as an ill-posed problem)
We want to formulate differentiation as an inverse problem. The direct problem formulation is as follows: Given u, find a function v with $v(0) = 0$ that satisfies

$$v(x) = \int_0^x u(t) \, \mathrm{d}t \quad \text{for} \quad t \in [0,1].$$

Accordingly, in the inverse problem we are given the function v which is continuously differentiable with $v(0) = 0$ and we want to determine $u = v'$. Thus, we have to solve the integral equation $Ku = v$, which leads to

$$Ku(x) := \int_0^x u(t) \, \mathrm{d}t, \quad t \in [0,1], \ u \in V := \left\{ v \in C^1 \,\middle|\, v(0) = 0 \right\}$$

with $K \in \mathcal{L}(C[0,1], C^1[0,1])$ and the equipped norm $\|\cdot\|_\infty$ on $C[0,1]$. It is:

$$\|Ku\|_\infty + \|(Ku)'\|_\infty = \sup_{x \in [0,1]} \left| \int_0^x u(t) \, \mathrm{d}t \right| + \sup_{x \in [0,1]} |u(x)|$$

$$\leq 2 \sup_{x \in [0,1]} |u(x)| = \|u(x)\|_\infty \,.$$

So, for $v \in C^1[0,1]$ with $v(0) = 0$ there exists a unique solution of $Ku = v$, namely $u = v'$. Especially, the inverse K^{-1} exists and is continuous. Furthermore, it is

$$\|K^{-1}v\|_\infty = \|v'\|_\infty \leq \|v\|_\infty + \|v'\|_\infty \quad \Rightarrow \quad \|K^{-1}\|_{C^1 \to C} \leq 1.$$

But in inverse problems the data is usually perturbed, thus, the vector we have at hand is not v but rather $v^\eta = v + \eta$. There are problems if $\eta(0) \neq 0$ or if η is not differentiable. One way out of this situation is to allow bounded errors, i.e. $K \in \mathcal{L}(L^\infty(0,1), L^\infty(0,1))$. Then, K is no longer surjective and in general the inverse K^{-1} does not exist. In cases of the differentiation, the inverse is not continuous. For $v \in C^1[0,1]$ with $v(0) = 0$ and $k \in \mathbb{N}$ we set

$$v^\eta(x) = v(x) + \eta \sin(kx).$$

A solution of $Ku^\eta = v^\eta$ is

$$u^\eta(x) = (v^\eta)'(x) = v'(x) + \eta k \cos(kx).$$

On the one hand we have $\|v^\eta - v\|_\infty = \eta$. On the other hand it holds $\|K^{-1}(v^\eta - v)\|_\infty = \eta k$. With $\eta_k = k^{-1/2}$ we get

$$\lim_{k \to 0} \|v^{\eta_k} - v\|_\infty = 0, \quad \text{but} \quad \lim_{k \to 0} \|K^{-1}(v^{\eta_k} - v)\|_\infty = \infty.$$

In Example 2.57 we saw that a small perturbation in the data can result in arbitrary large reconstruction errors. To overcome this issue we need more information on the initial data or additional assumptions on the solution. One way to do so and to overcome ill-posed minimization problems is to add a regularization term to the energy term.

In 1977, Tikhonov and Arsenin [TA77] proposed the consideration of the following problem:

$$\min_u \mathcal{F}(u) = \min_u \frac{1}{2} \int_\Omega |Ku - u_0|^2 \, dx + \frac{\lambda}{2} \int_\Omega |\nabla u|^2 \, dx. \tag{2.17}$$

The first term is the data fidelity term. In general we can consider a functional Φ, which describes the noise and is small only if the power of noise is small. The second term is a smoothing term. We make the assumption that there is some kind of spatial structure in measured datasets and images. Hence, considering neighborhood information seems sensible in order to obtain a "natural" approximation of $u^\dagger$. The parameter λ is a regularization parameter which weights how "smooth" the result u should be. In general the regularized minimization problem is of the form

$$\min_u \mathcal{F} = \min_u \Phi(Ku - u_0) + \lambda \Psi(u). \tag{2.18}$$

Therefore, a good choice of $\mathcal{F}$ should be a functional that simultaneously ensures some spatial regularity and preserve salient edges.

Following the examples in [CNCP10], there are few candidates for the choice of Ψ. Considering the minimization problem

$$\min_u \int_\Omega |u - u_0|^2 \, dx + \lambda \Psi(u) \tag{2.19}$$

we can set $\Psi(u) = \frac{1}{2} \int_\Omega |u|^2 \, dx$ or $\Psi(u) = \frac{1}{2} \int_\Omega |\nabla u|^2 \, dx$. The Euler-Lagrange equation for the latter case is

$$-\lambda \Delta u + u - u_0 = 0$$

where $\Delta u = \sum_k \partial^2 u / \partial x_k^2$ is the Laplacian of u.

Both terms are well-defined in $H^1(\Omega) = W^{1,2}(\Omega)$. The problem is that such functions cannot present discontinuities across hypersurfaces. In [CNCP10] they considered a small example. They considered a function $u : [0,1] \to \mathbb{R}$ which is a 1-dimensional function in $H^1(0,1)$. This u would have to be $1/2$-Hölder continuous, since for each $0 < s < t < 1$ it holds that

$$|u(t) - u(s)| \le |t - s|^{1/2} \|u\|_{H^1}. \tag{2.20}$$

In the two dimensional case it can be shown for almost every $y \in (0,1)$, $x \mapsto u(x,y) \in H^1(0,1)$ that u will be $1/2$-Hölder continuous in x. However, this means that the function cannot jump across vertical edges in an image.

Indeed, for $\Psi(u) = \frac{1}{2} \int_\Omega |\nabla u|^2 \, dx$ there is too much spatial regularization since the Laplacian has very strong isotropic smoothing properties. Hence, it does not preserve edges.

There are various approaches to find a functional that ensures some spatial regularity but preserves edges. There is a work by D. Geman and S. Geman [GG84] in the Bayesian context for one such approach.

In the continuous setting there are studies about the famous "Mumford-Shah" functional, cf. [MS89, AFP00].

However, this is particularly difficult to mathematically analyze and it is numerically complicated, since a non-convex minimization problem is solved there. In general we do not know whether or not a solution of this problem is a minimizer.

The most efficient methods to solve such problems are either stochastic algorithms or variational approximations; we will consider the latter.

In the previous section we saw a few approaches to formulate a minimization problem and a few candidates for functionals. But let us first take a step back and look at the energy term itself and the properties which it should have in order to provide the existence of minimizers. The following proof technique is known as the direct method of variational calculus.

Theorem 2.58 (Direct method of variational calculus)
Let X be a reflexive Banach space and $\mathcal{F} : X \to \overline{\mathbb{R}}$ be bounded from below, proper, coercive, and weakly lower semi-continuous. Then, the minimization problem

$$\min_{u \in X} \mathcal{F}(u)$$

has a solution $\hat{u} \in \mathrm{dom}\,\mathcal{F}$.

Proof. The direct method of variational calculus can be divided into the following steps:

1. Show the existence of a minimizing sequence $(u_n)_{n\in\mathbb{N}}$.
   $\mathcal{F}$ is proper and bounded from below, hence, there exists a $c \in \mathbb{R}$ with $c := \inf_{u\in X} \mathcal{F}(u)$. By the definition of the infimum we are able to construct a sequence which converges to c, namely $(y_n)_{n\in\mathbb{N}} \subset \mathrm{ran}\,\mathcal{F} \setminus \{+\infty\} \subset \mathbb{R}$. Therefore, there exists a sequence $(u_n)_{n\in\mathbb{N}} \subset X$ with $\mathcal{F}(u_n) \to c$ as $n \to \infty$. So, $(u_n)_{n\in\mathbb{N}}$ is a minimizing sequence. (Note, that this sequence does not have to be convergent.)

2. Show the existence of a weakly convergent subsequence $(u_{n_k})_{k\in\mathbb{N}}$.
   $\mathcal{F}$ is coercive and $\mathcal{F}(u_n) \to c < \infty$, $n \to \infty$. Therefore, $\|u_n\|_X \nrightarrow \infty$, $n \to \infty$ and the sequence is bounded. There exists a weakly convergent subsequence $(u_{n_k})_{k\in\mathbb{N}}$ with $u_{n_k} \rightharpoonup \hat{u} \in X$, $k \to \infty$. This weak limit is a candidate for a minimizer.

3. Show that $\lim_{k\to\infty} u_{n_k} = \hat{u}$ is a minimizer.
   We have $\mathcal{F}(u_{n_k}) \to c = \inf_{u\in X} \mathcal{F}(u)$ and with the weakly lower semi-continuity it follows that

$$c = \inf_{u \in X} \mathcal{F}(u) \leq \mathcal{F}(\hat{u}) \leq \liminf_{k \to \infty} \mathcal{F}(u_{n_k}) \leq c = \inf_{u \in X} \mathcal{F}(u).$$

Therefore, $\inf_{u\in X} \mathcal{F}(u) = \mathcal{F}(\hat{u})$ and $\hat{u} \in \mathrm{dom}\,\mathcal{F}$ is a minimizer of $\mathcal{F}$.

□

The following theorem gives another way to make sure that the minimization problem has a minimizer.

Theorem 2.59 (Subdifferential calculus and optimality)
Let $\mathcal{F} : X \to \overline{\mathbb{R}}$ be a proper and convex functional. Then, $u \in X$ is a minimizer of $\mathcal{F}$ if and only if $0 \in \partial \mathcal{F}(u)$.

Proof. For $0 \in \partial \mathcal{F}(u)$ the inequality for the subdifferential is

$$\mathcal{F}(u) + \langle 0\,, y - u \rangle_X \leq \mathcal{F}(y) \quad \forall y \in X.$$

Since $\langle 0\,, y - u \rangle_X = 0$, it follows that $\mathcal{F}(u) \leq \mathcal{F}(y)$ for all $y \in X$. Therefore, u is a minimizer of $\mathcal{F}$. On the other hand consider $0 \notin \partial \mathcal{F}(u)$. Then, there exists a $y \in X$ such that

$$\mathcal{F}(u) + \langle 0\,, y - u \rangle_X > \mathcal{F}(y),$$

Thus, $\mathcal{F}(u) > \mathcal{F}(y)$ and u is not a minimizer of $\mathcal{F}$. $\square$

Considering Equation (2.18) we can set $\Phi(Ku - u_0) = \frac{1}{2} \|u - u_0\|_X^2$ and $\lambda \Psi = F$ with certain conditions for F.

Lemma 2.60
Let $F : X \to \bar{\mathbb{R}}$ be proper, convex, and lower semi-continuous, and let $u_0 \in X$ be given. Then $\mathcal{F} : X \to \bar{\mathbb{R}}$ with

$$\mathcal{F}(u) = \frac{1}{2} \|u - u_0\|_X^2 + F(u)$$

is coercive.

Let us consider another candidate for a regularized minimization problem. In 1992, Rudin, Osher and Fatemi introduced the total variation as a regularizer [ROF92],

$$\min_u \frac{1}{2} \int_\Omega |u - u_0|^2 \, \mathrm{d}x + \lambda \, \mathrm{TV}(u). \tag{2.21}$$

This problem can also be written in terms of norms:

$$\min_u \frac{1}{2} \|u - u_0\|_2^2 + \lambda \, \||\nabla u|\|_1 \,.$$

That problem will be considered later on.

2.5 Solution methods for convex problems

In the previous sections we recalled convex analysis and variational methods. We also looked at the relationship between convex functionals and minimization problems according to the existence of minimizers.

Now, we will consider such convex optimization problems and recall a first-order primal-dual algorithm for this class of problems, proposed by A. Chambolle and T. Pock in 2011 [CP11].

We will state the primal-dual algorithm itself and introduce the resolvent of a monotone operator, additionally we will define the proximal mapping with a few examples for given operators that will be used in the following chapters.

Consider the general problem: Given are two finite-dimensional real vector spaces X and Y. These spaces are equipped with inner products $\langle \cdot \,, \cdot \rangle$ and norms $\| \cdot \| = \langle \cdot \,, \cdot \rangle^{1/2}$. Although one could consider the spaces to be general real separable Hilbert spaces.

In the general problem we will consider the generic saddle-point problem

$$\min_{x \in X} \max_{y \in Y} \langle Kx\,, y \rangle + F(x) - G^*(y) \tag{2.22}$$

where $F : X \to [0, \infty]$ and $G^* : Y \to [0, \infty]$ are proper, convex, and lower semi-continuous functions. G^* is the convex conjugate of a convex, lower semi-continuous function G, and K is a linear and continuous operator.

This saddle-point problem is the primal-dual formulation of the nonlinear *primal problem*

$$\min_{x \in X} F(x) + G(Kx) \qquad (2.23)$$

or of the corresponding *dual problem*

$$\max_{y \in Y} -F^*(-K^*y) - G^*(y). \qquad (2.24)$$

We refer to [Roc70] for more details.

Assume that this problem has at least one solution, namely the pair $(\hat{x}, \hat{y}) \in X \times Y$ which satisfies

$$K\hat{x} \in \partial G^*(\hat{y})$$
$$\text{and} \quad -K^*\hat{y} \in \partial F(\hat{x}), \qquad (2.25)$$

where ∂F and ∂G^* are the subgradients of F and G^*, respectively.

Because of the connection of the minimizer and the subgradient we will take a closer look at the mapping $x \mapsto \partial F(x)$ where $F : X \to \mathbb{R}$ is again a convex functional.

Consider the following functional for $z \in X$:

$$\mathcal{F} : X \to \mathbb{R}, \quad \mathcal{F}(x) = \frac{1}{2} \|x - z\|_X^2 + F(x). \qquad (2.26)$$

In the following, let I be the identity.

Lemma 2.61
Let $F : X \to \mathbb{R}$ be proper, convex and lower semi-continuous. Then, $I + \partial F$ is surjective.

Proof. We consider Functional (2.26). Since $\| \cdot \|_X$ and F are proper, convex, and lower semi-continuous function the functional $\mathcal{F}$ is proper, convex, and lower semi-continuous. It can also be shown that $\mathcal{F}$ is coercive. We want to estimate a lower bound of $\mathcal{F}(x)$ which contains fixed terms and $\|x\|_X$. For this, we will use that F is bounded from below and use the reverse triangle inequality. Hence, there exists a unique $\hat{x} \in X$ such that

$$\mathcal{F}(\hat{x}) = \min_{x \in X} \mathcal{F}(x).$$

Thus, $0 \subset \partial\mathcal{F}(\hat{x})$. For $z \subset X$ the subgradient of $\mathcal{F}$ is given by $\partial\mathcal{F}(\hat{x}) = \hat{x} - z + \partial F(\hat{x})$. Together we get

$$0 \in \partial\mathcal{F}(\hat{x}) \quad \Leftrightarrow \quad 0 \in \hat{x} - z + \partial F(\hat{x})$$
$$\Leftrightarrow \quad z \in (I + \partial F)(\hat{x})$$

which gives the surjectivity of $I + \partial F$. $\qquad \square$

Definition 2.62 (Monotone maps)
A multivalued map $F : X \rightrightarrows X$ on a Hilbert space X is called *monotone* if for $y_1 \in F(x_1)$ and $y_2 \in F(x_2)$ it holds that

$$\langle y_1 - y_2, \, x_1 - x_2 \rangle_X \geq 0.$$

A multivalued map is called *maximally monotone* if its graph $F \subset X \times X$ is not properly contained in another subset.

Clearly, the identity mapping $x \mapsto \{x\}$ is monotone. Also, if $F : X \to \mathbb{R}$ is convex, $\partial F : X \rightrightarrows X$, $x \mapsto \partial F(x)$ is monotone.

For completeness, we give the generalization of Lemma 2.61.

Theorem 2.63 (Minty's theorem [Min62])
Let $F : X \rightrightarrows X$ be monotone. Then, F is maximally monotone if and only if $I+F$ is surjective.

Corollary 2.64
Let $F : X \to \mathbb{R}$ be proper, convex, and lower semi-continuous. Then $\partial F : X \to X$ is maximally monotone.

Now, we know that $I + \partial F$ is surjective for any F which is proper, convex, and lower semi-continuous and the strict convexity of J shows that $(I + \partial F)^{-1}$ is single-valued even if ∂F is not. Instead of using ∂F itself for an algorithm we can use this maximally monotone operator. Thus, we define the resolvent and the proximal point operator as follows.

Definition 2.65
Let $F : X \rightrightarrows X$ be a maximally monotone. The *resolvent* is defined as
$$\mathcal{R}_F : X \to X, \quad \mathcal{R}_F(x) = (I + F)^{-1}(x).$$

Definition 2.66
Let $F : X \to \bar{\mathbb{R}}$ be proper, convex, and lower semi-continuous. The *proximal operator* is defined as
$$\mathrm{prox}_F : X \to X, \quad \mathrm{prox}_F(z) = \arg\min_{x \in X} \frac{1}{2}\|x - z\|_X^2 + F(x).$$

Since ∂F is maximally monotone it follows that
$$\mathrm{prox}_F = (I + \partial F)^{-1} = \mathcal{R}_{\partial F}. \tag{2.27}$$

The $x = \mathcal{R}_{\partial F}(z)$ is called *proximal point*.

Lemma 2.67
Let $F : X \to \bar{\mathbb{R}}$ be proper, convex, and lower semi-continuous and let $x, y \in X$. Then, for any $\gamma > 0$ it holds that
$$y \in \partial F(x) \quad \Leftrightarrow \quad x = \mathrm{prox}_{\gamma F}(x + \gamma y).$$

Proof. Taking the left hand side of the statement we get
$$\begin{aligned} y \in \partial F(x) \quad &\Leftrightarrow \quad x + \gamma y \in x + \gamma \partial F(x) \\ &\Leftrightarrow \quad x + \gamma y \in (I + \gamma \partial F)(x) \\ &\Leftrightarrow \quad x \in (I + \gamma \partial F)^{-1}(x + \gamma y). \end{aligned}$$

With $\gamma \partial F = \partial(\gamma F)$ it follows
$$x = \mathrm{prox}_{\gamma F}(x + \gamma y)$$

and thus, $\mathrm{prox}_{\gamma F} = \mathcal{R}_{\gamma \partial F}$. □

Corollary 2.68
Let $F : X \to \bar{\mathbb{R}}$ be proper, convex, and lower semi-continuous and let $\gamma > 0$. Then, $\hat{x} \in \mathrm{dom}(F)$ is a minimizer of F if and only if
$$\hat{x} = \mathrm{prox}_{\gamma F}(\hat{x}).$$

Theorem 2.69 (Moreau's decomposition)
Let $F : X \to \bar{\mathbb{R}}$ be proper, convex, and lower semi-continuous. Then, for all $x \in X$ it holds that

$$x = \operatorname{prox}_F(x) + \operatorname{prox}_{F^*}(x). \tag{2.28}$$

Proof. Let $w = \operatorname{prox}_F(x)$. With Theorems 2.46 and 2.67 we obtain:

$$
\begin{aligned}
w = \operatorname{prox}_F(x) = \operatorname{prox}_F(w + (x - w)) \quad &\Leftrightarrow \quad x - w \in \partial F(w) \\
&\Leftrightarrow \quad w \in \partial F^*(x - w) \\
&\Leftrightarrow \quad x - w = \operatorname{prox}_{F^*}((x - w) + w) = \operatorname{prox}_{F^*}(x).
\end{aligned}
$$

$\square$

Lemma 2.70
Let $F : X \to \bar{\mathbb{R}}$ be proper, convex, and lower semi-continuous.

1. Let $\gamma \neq 0$, $z \in X$, and $H(x) = F(\gamma x + z)$. Then,

$$\operatorname{prox}_H(x) = \gamma^{-1}\left(\operatorname{prox}_{\gamma^2 F}(\gamma x + z) - z\right).$$

2. For $\gamma > 0$ it holds that

$$\operatorname{prox}_{\gamma F^*}(x) = x - \operatorname{prox}_{\gamma^{-1}F}(\gamma^{-1}x).$$

3. Let $G : Y \to \bar{\mathbb{R}}$ be proper, convex, and lower semi-continuous, $\gamma > 0$ and $H(x, y) = F(x) + G(y)$. Then,

$$\operatorname{prox}_{\gamma H}(x, y) = \begin{pmatrix} \operatorname{prox}_{\gamma F}(x) \\ \operatorname{prox}_{\gamma G}(x) \end{pmatrix}.$$

Proof. 1. Let $H(x) = F(\gamma x + z)$ and $\gamma > 0$. Then,

$$
\begin{aligned}
\hat{w} = \operatorname{prox}_H(x) &= \arg\min_{w \in X} \; \frac{1}{2}\|x - w\|_X^2 + H(w) \\
&= \arg\min_{w \in X} \; \frac{1}{2}\|x - w\|_X^2 + F(\gamma w + z) \\
\Leftrightarrow \quad \hat{v} &= \arg\min_{v \in X} \; \frac{1}{2}\left\|x - \gamma^{-1}(v - z)\right\|_X^2 + F(v) \\
&= \arg\min_{v \in X} \; \frac{1}{2\gamma^2}\|(\gamma x + z) - v\|_X^2 + F(v) \\
&= \arg\min_{v \in X} \; \frac{1}{2}\|(\gamma x + z) - v\|_X^2 + \gamma^2 F(v) \\
&= \operatorname{prox}_{\gamma^2 F}(\gamma x + z) \\
\Leftrightarrow \quad \hat{w} &= \gamma^{-1}(\hat{v} - z) = \gamma^{-1}(\operatorname{prox}_{\gamma^2 F}(\gamma x + z) - z).
\end{aligned}
$$

2. Let $y = \text{prox}_{\gamma F^*}(x)$. Then,

$$x \in (I + \gamma \partial F^*)(y)$$
$$\Leftrightarrow \quad y \in (I + \gamma \partial F^*)^{-1}(x)$$
$$\Leftrightarrow \quad \frac{x - y}{\gamma} \in \partial F^*(y)$$
$$\Leftrightarrow \quad y \in \partial F\left(\frac{x - y}{\gamma}\right)$$
$$\Leftrightarrow \quad x \in \gamma^{-1}\partial F\left(\frac{x - y}{\gamma}\right) + \frac{x - y}{\gamma}$$
$$\Leftrightarrow \quad \frac{x - y}{\gamma} \in (I + \gamma^{-1}\partial F)^{-1}\left(\frac{x}{\gamma}\right)$$
$$\Leftrightarrow \quad y \in x - \gamma^{-1}(I + \gamma^{-1}\partial F)^{-1}\left(\frac{x}{\gamma}\right)$$
$$\Leftrightarrow \quad \text{prox}_{\gamma F^*}(x) = x - \gamma^{-1}\,\text{prox}_{\gamma^{-1}F}\left(\frac{x}{\gamma}\right).$$

3. Let $(\hat{v}, \hat{w}) = \text{prox}_{\gamma H}(x, y)$. Then,

$$\begin{pmatrix} \hat{v} \\ \hat{w} \end{pmatrix} = \text{prox}_{\gamma H}(x, y) = \underset{(v,w) \in X \times Y}{\arg\min} \ \frac{1}{2}\|(v, w) - (x, y)\|_{X \times Y}^2 + \gamma H(v, w)$$

$$= \underset{(v,w) \in X \times Y}{\arg\min} \ \frac{1}{2}\|v - x\|_X^2 + \gamma F(v) + \|w - y\|_Y^2 + \gamma G(w)$$

$$= \begin{pmatrix} \text{prox}_{\gamma F}(x) \\ \text{prox}_{\gamma G}(y) \end{pmatrix},$$

since the arguments can be divided into components in terms of X and in terms of Y.

$\square$

In the following chapters the proximal mappings of the indicator function and a norm function will frequently be used.

Example 2.71

1. Let $F(x) = \mathcal{I}_C(x)$ for a convex set $C \subset X$. Then, the proximal operator is equal to a projection onto the set C:

$$\text{prox}_F(x) = \underset{y \in X}{\arg\min} \ \frac{1}{2}\|x - y\|_X^2 + F(y) = \underset{y \in C}{\arg\min} \ \frac{1}{2}\|x - y\|_X^2 = \text{proj}_C(x).$$

2. Let $F(x) = \|x\|_X$. In Example 2.48 we have seen that for the dual function holds: $F^*(x) = \mathcal{I}_{\|\cdot\|_{X^*} \leq 1}(x)$. With Moreau's decomposition (2.28) and the previous example it yields

$$\text{prox}_F(x) = x - \text{prox}_{F^*}(x) = x - \text{proj}_{\|\cdot\|_{X^*} \leq 1}(x).$$

If the resolvent operator, i.e. the proximal mapping has a closed-form representation or can efficiently be solved up to a high precision, for example by Newton's method, we are able to use the following algorithm (see [CP11]) for problems like (2.22).

How the parameters within the initialization have to be chosen will be discussed later on.

Algorithm 1: Chambolle-Pock's primal-dual algorithm [CP11].

Initialization: Choose $\tau, \sigma > 0$, $\theta \in [0, 1]$, $(x^0, y^0) \in X \times Y$ and set $\bar{x}^0 = x^0$.
Iterations $(n \geq 0)$: Update $x^n, y^n, \bar{x}^n$ as follows:

$$\begin{cases} y^{n+1} = (I + \sigma \partial G^*)^{-1}(y^n + \sigma K \bar{x}^n) \\ x^{n+1} = (I + \tau \partial F)^{-1}(x^n - \tau K^* y^{n+1}) \\ \bar{x}^{n+1} = x^{n+1} + \theta(x^{n+1} - x^n) \end{cases} \qquad (2.29)$$

DEFLECTOMETRIC MEASUREMENT AND OPTIMIZATION OF DATASETS

In this chapter we describe a collaboration with the Institute of Production Measurement Technology (IPROM) at the Technical University Braunschweig. The task at hand is an inverse problem for a certain application in deflectometric measurement.

Let us give a brief overview of the problem before we go into details in the next sections. Let a dataset be given which describes a measured surface. For this specific application we consider specular objects such as small mirrors. The dataset consists of a set of three dimensional surface coordinates and a separate set of surface normals. The measurement process which we will look at later on gives these two different sets as independent outputs. The problem is that the surface coordinates and normals do not have the same accuracy. More precisely, the accuracy of the normal vectors is three orders of magnitude higher than that of the point coordinates, since the direct measurement of these points is more sensitive to noise. (For more details on that topic, as well as the specific technical details of this measurement method, see [Pet04, Pet06]). The different accuracies lead to inconsistencies in the dataset.

Now, the goal is to use the high accuracy of the surface normals to increase the accuracy of the coordinates. In this chapter we will give a brief description of the geometry of this measurement technique and then formulate the data fusion problem mathematically. Further-more, we will discuss ideas of solution techniques and give an approach to denoise the surface coordinates with respect to the measured normal vectors, first for continuous and later for discontinuous surfaces. Additionally, we will give remarks about how to handle these datasets, since the surface coordinates do not lie on a regular grid, and some coordinates are missing from the output due to the measurement process itself.

All these will be tested with real measured datasets and be displayed later on in the results section, Section 3.5.2 (see also [KLF+14]).

3.1 Deflectometry

The term *deflectometry* denotes the measurement of a reflective object without actual contact. A measurement method like the one we will consider in this chapter uses at least one camera and one projector, in this case a monitor since we consider specular objects. The monitor projects a so-called fringe pattern on the surface of an object and its reflection will be recorded with the camera as a coded information. This setup is sketched in Figure 3.1.

In general we have three parts of the measurement setup. First the object itself: In case of deflectometry this could be a mirror or lens, for example. Throughout the measurement process

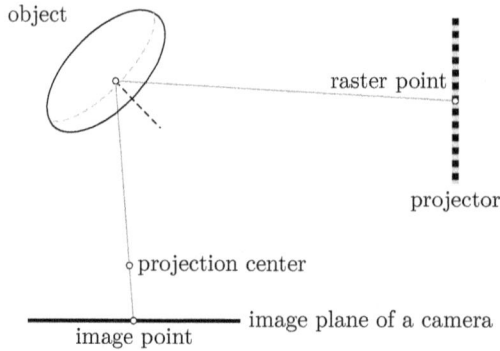

Figure 3.1: Measurement setup to measure an object (here, usually specular); the setup consists
of one camera and one projector.

the object is fixed in position. Second, the camera side: Ideally, for each point coordinate on
the surface of the object we are able to find one distinct light ray. That can be interpreted as a
straight line which goes through the projection center of the camera and the image point on the
camera image plane. And third, the projection side: There, a fringe pattern is projected onto
the object's surface. This pattern is usually a sine or cosine function. With only one projector
(or projector position) and an object fixed in position it is not possible to find a unique light
ray that corresponds to the light ray on the camera side, see Figure 3.2.

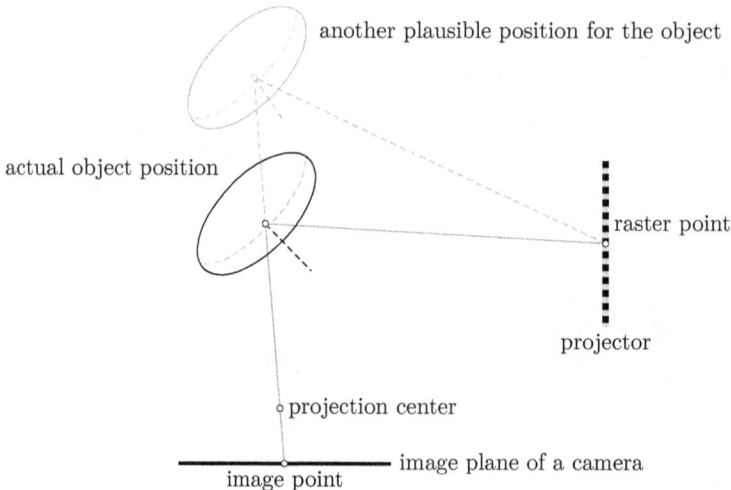

Figure 3.2: Light rays on the camera side are unique because these are going through the
projection center and a point on the image plane of the camera. Light rays on the
projection side are not unique with just one projector and object position. The
actual position, depicted in black below, cannot be determined by this setup.

There are different ways to solve this problem. One is to add a second camera (or camera position) to the measurement setup. The technique where a second camera position is
used is called *stereo vision* or *passive reflection grating photogrammetry*. With known po-

sitions of the cameras we are able to determine a unique light ray for each camera which intersects in the corresponding point on the object's surface. This can be realized if we know the *epipolar geometry* of the camera system and the reflection properties of the measured object (see [LRKB13, Luh00, Pet04, Pet06]).

Another approach is called *active reflection grating photogrammetry*. Here, an additional position of the projector is used instead of a second camera position. The idea is basically the same. By adding another reference system we are able to determine the position of a measured object in three-dimensional space. But instead of using the projector side just for its coding ability a second reference point on the projector side is added. These two points span a second unique straight line on the other side of the measurement system (see Figure 3.3(a)).

The projector is an LCD display which can be shifted in various positions. There are two CCD-cameras and in the center of the setup is a calibration mirror where the object will be in a fixed position during the measurement process. This setup (see Figure 3.3(b)) can be used for the passive as well as for the active reflection grating. (For further technical details see [Pet06].)

The two dashed lines on the right in Figure 3.3(b) represent two different projector positions. There are two light rays, one on each on the camera and on the projector side. From the intersection of these light rays one can calculate the bisection vector of these light rays which is the normal direction of the surface in this surface point.

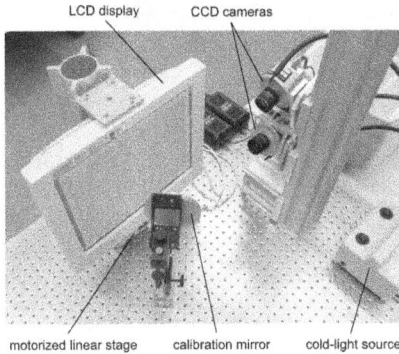

(a) Experimental setup of the reflection grating photogrammetry technique [Pet04, Pet06].

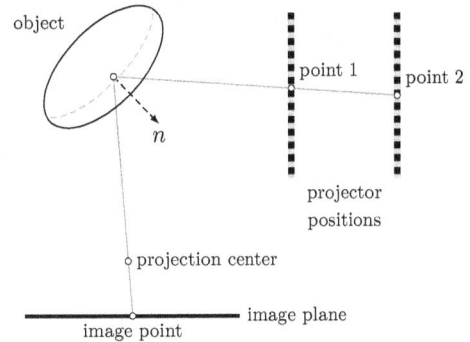

(b) Schematic sketch with two projector positions; points 1 and 2 determine a unique light ray on the projector side.

Figure 3.3: Setup of the reflective grating photogrammetry method.

For the measurement process, a fringe pattern, usually in form of a sine curve, are projected onto the object. Several phase shifted patterns are used for spatial coding (see Figure 3.4). This shall ensure that every measurable point on the surface has a unique reference pattern, i.e. a unique pixel on the monitor according to the phase of the fringe pattern. These pixels can be seen in Figure 3.3(b) denoted with *point 1* and *point 2*.

Now, we have a unique light ray for a point coordinate on the projector side which is spanned by the points 1 and 2 and a unique light ray on the camera side going through the projection center and the image point on the plane. The intersection of these light rays defines the location of the object in three dimensional space. (Usually there will be no intersection due to uncertainties in the measurement. So, one has to calculate an approximation.) Additionally,

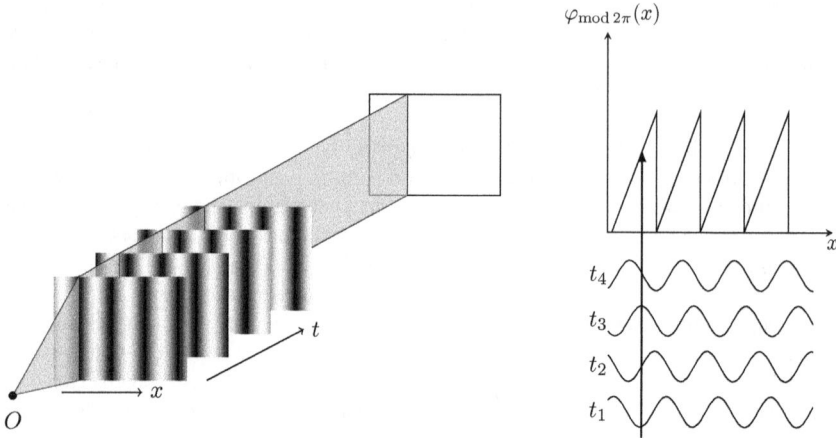

Figure 3.4: Phase shift method for object measurement. Over time t sine functions with the same period length are projected onto the object. In this example four sine functions are used.

the orientation of the surface point can be determined, i.e., the surface normal vectors, by the law of reflection.

A closer investigation of the measurement data reveals that this direct measurement of coordinates is more sensitive to noise than the measurement of the surface normals. It is shown by M. Petz [Pet06] that the noise of the measured surface coordinates is approximately three orders of magnitude higher than it is for the reconstruction based on the integration of the surface slopes (see also [KLF$^+$14]).

Since the surface normals are calculated based on integration of the surface slopes, the normals are not that sensitive to noise and are more accurate. So, the measured dataset suffers from inconsistency. This means the measured surface points do not accurately describe the object.

In the next subsections we develop ideas for data fusion techniques such that after optimization the resulting dataset will be more consistent. In order to obtain reasonable results we have to consider occurring measurement uncertainties and the exact position of each pixel before our optimization approaches. The coordinates of each pixel are disturbed by noise that follows a Gaussian distribution in the third direction with known standard deviation σ. Hence, the distribution of the sum of all squared distances in the third direction follows from a χ^2-distribution with the number of degrees of freedom equal to the number $\#\boldsymbol{P}$ of valid measured points. We deduce that the expected value of this sum is equal to $\#\boldsymbol{P}\sigma^2$. In the following approaches we will use $\delta = \sigma\sqrt{\#\boldsymbol{P}}$ as a tolerance value (or constraint parameter) for deviations of the measured point coordinates (cf. [KLF$^+$14]).

3.2 Mathematical Setup

We get a measured surface as triples of coordinates and additionally measured normal vectors from the active reflection grating photogrammetry measurement process. For further studies we consider the surface to be a graph over a domain $\Omega \subset \mathbb{R}^2$. Let

$$u^0 : \Omega \to \mathbb{R}$$

be this measured quantity and let $u^\dagger$ be the true surface graph with no noise or measurement errors. In Figure 3.5 we can see the schematic idea.

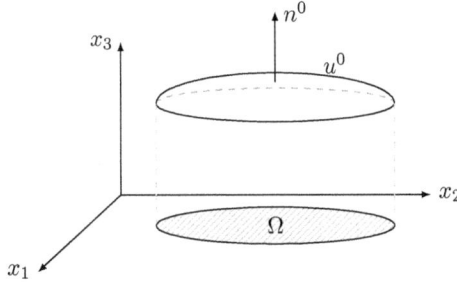

Figure 3.5: Domain Ω and u^0 as graph over the domain and one of the normal vectors. The camera would be in x_3 direction.

We also know that the coordinates of each pixel are disturbed by Gaussian noise, say η, in the third direction which is u^0 in this setup. Thus, the relation between u^0 and $u^\dagger$ is

$$u^0 = u^\dagger + \eta. \tag{3.1}$$

Further, we get a set of measured normal vectors which can be expressed as a mapping with the same domain as the surface. These we denote by

$$n^0 : \Omega \to \mathbb{R}^3.$$

Similar to images, where we know if a pixel is a neighbor of another pixel due to the given regular grid of an image, we get additional neighborhood information of each surface coordinate. For the domain Ω in the x_1-x_2-plane we define an underlying discrete grid with index i for the first direction and j for the second direction of the coordinate system. In each index point (i, j) of this grid lies a measured (x_1, x_2)-coordinate. The surface coordinates in the three-dimensional space are given by

$$p_{i,j} = \begin{pmatrix} x_1 \\ x_2 \\ u^0(x_1, x_2) \end{pmatrix}_{i,j} \in \mathbb{R}^3. \tag{3.2}$$

The neighborhood properties are encoded in the (i, j)-grid. The direct neighborhood is the set of pixels that differ by 1 in absolute value in the indices (cf. Figure 3.6). We label the neighborhood of a pixel $p_{i,j}$ with

$$\mathcal{N}(i,j) = \left\{ p \in \mathbb{R}^3 \,\middle|\, p \text{ is neighbor of } p_{i,j} \right\}. \tag{3.3}$$

We define the index set for neighboring points by

$$I_{i,j} := \left\{ (\hat{k}, \hat{\ell}) \in \{i-1, i, i+1\} \times \{j-1, j, j+1\} \,\middle|\, \begin{array}{l} (\hat{k}, \hat{\ell}) \neq (i, j), \\ 1 \leq \hat{k} \leq M, \\ 1 \leq \hat{\ell} \leq N \end{array} \right\}. \tag{3.4}$$

The whole setup of the point coordinates, its not-regular grid, and the regular index grid is sketched in Figure 3.6. Now, the setup is clear in a mathematical sense and we go on and work on the ideas of data fusion. In the next section we will formulate approaches based on the idea of using the measured normal vectors as a kind of fixed and trusted entity.

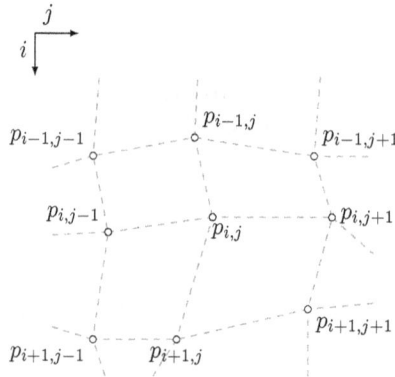

Figure 3.6: Neighborhood $\mathcal{N}(i,j)$ forming a not regular grid of pixels.

3.3 Data fusion approaches by usage of measured surface normals

In this section we will work on two approaches to fuse the measured data, i.e. the measured surface normal vectors and surface point coordinates, whereas the accuracy of the normals is three orders of magnitude higher (see also [Pet04, Pet06]).

First, we briefly discuss an approach which uses the normal vectors directly. And second, we discuss in detail an approach which uses the normal vectors to parametrize the surface through gradient information.

Let us start with the first idea. We take the surface normals n^0 as the "true" surface orientation and let them be fixed throughout the optimization process. From the point coordinates and their neighborhood information we can calculate an approximate normal vector $n_{i,j}$ for each coordinate by fitting a plane through all eight neighbors and the point coordinate itself (see Figure 3.6).

Alternatively, we calculate normal vectors via crossproducts of directional vectors pointing from the center point to each neighbor (see Figure 3.7). Here, one can take any two neighboring points to calculate $n_{i,j}$ or even more of these points. Then, we would have different directions and also calculate the normals as a mean of the $n_{i,j}$ according to two directions. With this we are able to formulate an optimization problem which minimizes the difference between the set of measured normals n^0 and the set of calculated normal vectors n. Taking all neighboring points to calculate the normals can lead to problems like when there is a discontinuity between two index pairs. The resulting normal in such a point would not point in the correct direction.

This approach is similar to an image denoising technique proposed by Lysaker et al. in 2004 [LOT04]. They use a total variation filter to first smooth the normal vectors of the levelsets of a given noisy image. Then they try to find a surface to fit the smoothed normal vectors. Here, we could just try to find this fitting surface directly since the measured normals are the trusted measure. But this is a highly non-linear problem (cf. [Kom13]).

Hence, we propose a slightly different approach by switching to surface gradients. Taking a step back and considering again the overall task, we see that a surface can be described not only by normal vectors but also by a parametric equation as

$$\varphi : \Omega \to \mathbb{R}^3, \quad \varphi(x_1, x_2) = \begin{pmatrix} x_1 \\ x_2 \\ u^0(x_1, x_2) \end{pmatrix}.$$

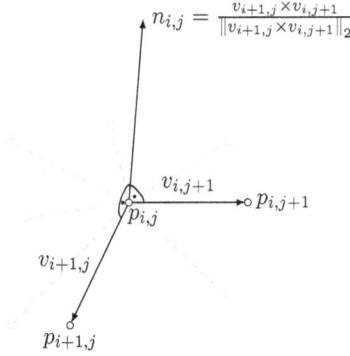

$$n_{i,j} = \frac{v_{i+1,j} \times v_{i,j+1}}{\|v_{i+1,j} \times v_{i,j+1}\|_2}$$

Figure 3.7: Given the neighborhood information of one pixel point $p_{i,j}$. One could consider all neighbors $p_{k,\ell} \in \mathcal{N}(i,j)$ (grey dashed lines) in order to calculate normal vectors $n_{i,j}$ according to the directional vectors pointing from $p_{i,j}$ to its neighbors. In this example only two neighbors are taken into consideration for $n_{i,j}$.

Considering this parametric formulation of the surface in the mathematical setup section we see

$$(\varphi(x_1, x_2))_{i,j} = p_{i,j}.$$

Hence, this parametric formulation of the surface describes every measured pixel given in the point cloud. Further, we know that the partial derivatives of such a parametric entity are tangential vectors to the surface. Therefore, we are able to calculate normal vectors by means of a crossproduct such as

$$n\left(\varphi(\Omega)\right)\big|_{(x_1,x_2)} = \begin{pmatrix} -\partial_1\, u^0(x_1, x_2) \\ -\partial_2\, u^0(x_1, x_2) \\ 1 \end{pmatrix} = \partial_1\, \varphi(x_1, x_2) \times \partial_2\, \varphi(x_1, x_2).$$

With this identification we are able to formulate the measured surface normals $n^0 = (n_1^0, n_2^0, n_3^0)^T$ by

$$\frac{1}{n_3^0}\, n^0 = \begin{pmatrix} n_1^0/n_3^0 \\ n_2^0/n_3^0 \\ 1 \end{pmatrix} = \begin{pmatrix} -\partial_1\, u^0 \\ -\partial_2\, u^0 \\ 1 \end{pmatrix} = \begin{pmatrix} -\nabla u^0 \\ 1 \end{pmatrix}.$$

Thus, we obtain measured gradient vectors

$$\nabla u^0(x_1, x_2) := -\begin{pmatrix} \frac{n_1^0(x_1,x_2)}{n_3^0(x_1,x_2)} \\ \frac{n_2^0(x_1,x_2)}{n_3^0(x_1,x_2)} \end{pmatrix}. \tag{3.5}$$

Note that the symbol ∇ is not the operator but we define a fixed calculated gradient ∇u^0 such that this is the whole variable. Now we can formulate an optimization functional. For example we formulate a least-squares problem which considers the gradient of u^0 as reference value

$$\min_u \frac{1}{2}\, \|\nabla u^0 - \nabla u\|_{L^2(\Omega)}^2 = \min_u \frac{1}{2} \int_\Omega \left|\nabla u^0(x_1, x_2) - \nabla u(x_1, x_2)\right|^2 \, d(x_1, x_2) \tag{3.6}$$

where $|\cdot|$ is the 2-norm. It is clear that u^0 would solve this minimization problem with 0 objective. Since we want to calculate more accurate surface coordinates, we will add a constraint later on.

Before we can discuss how to handle this problem according to the choice of norm and solution methods we have to find a way to calculate a gradient ∇u with respect to the measured point coordinates.

If the point coordinates lie on a regular grid we could simply calculate finite differences. But the coordinates do not lie on a regular grid, see again Figure 3.6. However, neighborhood information is available through the measurement process. Thus, we are able to express directional derivatives for each point coordinate. We realize this in two ways in order to generate a linear system which we can solve for ∇u.

In general the definition of a directional derivatives of a function $u : \Omega \to \mathbb{R}$ in some direction $v \in \mathbb{R}^2$ is defined by

$$\lim_{t \to 0} \frac{u(x + tv) - u(x)}{t\,\|v\|}. \tag{3.7}$$

In case of a differentiable function u we are able to write the directional derivative with the inner product

$$\partial_v u = \langle v\,,\, \nabla u \rangle. \tag{3.8}$$

Equation 3.8 above gives us a way to generate a linear system for ∇u as follows.

First, we consider the directional derivative in a single point of u^0, i.e. we look at the underlying grid and take $u_{i,j}^0(x_1, x_2)$. The neighborhood properties are known, so we can use them to determine directions for further calculations. For better approximation of the derivative we take all pixels $p \in \mathcal{N}(i, j)$ (cf. Equation (3.3)) into account (see also Figure 3.8). Hence, the first formulation for the directional derivative is given by

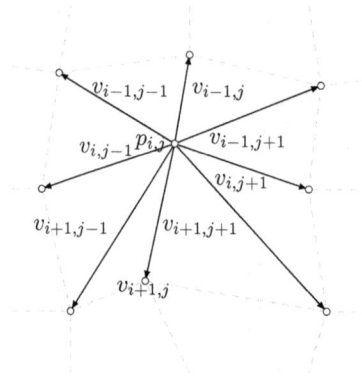

Figure 3.8: Directional vectors $v_{k,l}$, $(k, l) \in I_{i,j}$ in the neighborhood of a pixel $p_{i,j}$. There are at most eight direct neighbors of one pixel.

$$\partial_{v_{k,\ell}} u_{i,j}^0 = \left\langle v_{k,\ell}\,,\, \left(\nabla u^0\right)_{i,j} \right\rangle. \tag{3.9}$$

We still need a second formulation for the directional derivatives in order to formulate a linear system. Here, we can use finite differences. We will also note a few properties that will be helpful for the implementation.

We look at a single coordinate $u_{i,j}^0$ of $u \in \mathbb{R}^{M \times N}$. Consider all $p_{k,\ell} \in \mathcal{N}(i,j)$ with $(k,\ell) \in I_{i,j}$. The directional derivative in corresponding direction $v_{k,\ell}(x_1, x_2)$ with $(k,\ell) \in I_{i,j}$, is given by

$$\partial_{v_{k,\ell}} u_{i,j}^0(x_1, x_2) = u_{k,\ell}^0(x_1, x_2) - u_{i,j}^0(x_1, x_2),$$

where the directional vectors are defined as

$$v_{k,\ell} = \begin{pmatrix} (x_1)_{k,\ell} - (x_1)_{i,j} \\ (x_2)_{k,\ell} - (x_2)_{i,j} \end{pmatrix}. \tag{3.10}$$

We now drop the coordinates (x_1, x_2) to simplify the notation. Then, we obtain an equation for $(\nabla u^0)_{i,j}$ by identifying the analytical description of the directional derivative with the finite difference representative

$$\left\langle v_{k,\ell}, \left(\nabla u^0\right)_{i,j} \right\rangle = u_{k,\ell}^0 - u_{i,j}^0.$$

Note that we only calculate the finite difference representation of directional derivatives and use these and the formulation via inner product to solve for ∇u^0.

Taking all possible directions $v_{k,\ell}$ of one single pixel $p_{i,j}$, we write them into one matrix $V_{i,j} \in \mathbb{R}^{\#\mathcal{N}(i,j) \times 2}$ and form an over-determined linear system for $(\nabla u^0)_{i,j}$:

$$V_{i,j}\left(\nabla u^0\right)_{i,j} = \begin{pmatrix} v_{i-1,j-1}^T \\ \vdots \\ v_{i+1,j+1}^T \end{pmatrix} \left(\nabla u^0\right)_{i,j} = \begin{pmatrix} u_{i-1,j-1}^0 - u_{i,j}^0 \\ \vdots \\ u_{i+1,j+1}^0 - u_{i,j}^0 \end{pmatrix}.$$

In order to solve this equation for $(\nabla u^0)_{i,j}$ we use the pseudo-inverse of $V_{k,\ell}$ which is a generalization of an inverse matrix.

Definition 3.1 (Pseudo-inverse)
Let $A \in \mathbb{K}^{m \times n}$. The *pseudo-inverse* of A is a matrix $A^+ \in \mathbb{K}^{n \times m}$ enjoying the following properties:

1. $AA^+A = A$ and $A^+AA^+ = A^+$,
2. $(A^+A)^* = A^+A$ and $(AA^+)^* = AA^+$ (Hermitian).

Now, we can write down a linear equation for any index pair (i,j), i.e. the gradient in one point,

$$\left(\nabla u^0\right)_{i,j} = V_{i,j}^+ \begin{pmatrix} u_{i-1,j-1}^0 - u_{i,j}^0 \\ \vdots \\ u_{i+1,j+1}^0 - u_{i,j}^0 \end{pmatrix}$$

$$= V_{i,j}^! \underbrace{\begin{pmatrix} 1 & 0 & 0 & 0 & -1 & 0 & 0 & 0 & 0 \\ 0 & 1 & 0 & 0 & -1 & 0 & 0 & 0 & 0 \\ & & & & \vdots & & & & \\ 0 & 0 & 0 & 0 & -1 & 0 & 0 & 0 & 1 \end{pmatrix}}_{=:S} \begin{pmatrix} u_{i-1,j-1}^0 \\ \vdots \\ u_{i,j}^0 \\ \vdots \\ u_{i+1,j+1}^0 \end{pmatrix}. \tag{3.11}$$

The next step is to expand this to obtain a linear equation for the whole gradient of u. The scheme for the finite differences on the right hand side in Equation (3.11) can be written with a sparse matrix S such that $(Su^0)_{i,j}$ is the vector containing the finite differences of $u_{i,j}^0$ in all possible directions (as already done in Equation (3.11)). For each pixel we calculate the product $V_{i,j}^+ S$ and formulate a differential operator $D \in \mathcal{L}(\mathbb{R}^{\#P}, \mathbb{R}^{\#P \times 2})$ where $\#P$ is the number of *valid* measured points.

In practice the dataset is not necessarily a full dataset which means that due to the measurement process values at some index point (i, j) can be missing. Thus, we need to be careful within the implementation and consider only the valid measured points when taking finite differences, as the number of neighbors may be decreased. For the implementation we formulate D as a sparse matrix which has the general form that the upper half will be corresponding to the partial derivatives in first direction, i.e. ∂_1, and the lower half will be corresponding to the second direction, i.e. ∂_2. Because of the calculated finite differences with respect to the 8-neighborhood of a pixel, the resulting matrix will have the form of three tridiagonal lines. The linear operator $D : \mathbb{R}^{\#P} \to \mathbb{R}^{\#P \times 2}$ is defined in a way such that schematically

$$(Du)_{\cdot,1} = \begin{bmatrix} \end{bmatrix} u \qquad \text{and} \qquad (Du)_{\cdot,2} = \begin{bmatrix} \end{bmatrix} u$$

and this yields

$$\nabla u = Du.$$

Note, that D is calculated before the minimization and is fixed during the optimization process.

With this, the resulting minimization functional can be written as

$$\min_u \frac{1}{2} \| Du - \nabla u^0 \|_{L^2(\Omega)}^2.$$

Now, we can consider the image model (3.1) and we have $\delta = \sigma\sqrt{\#P}$ as a tolerance value for the deviations of the measured point coordinates, where σ is the standard deviation of the additive noise η and $\#P$ is the number of valid point coordinates. (This topic will be revised in Chapter 5, Section 5.5. There, we will show exactly the statement made above.)

Thus, we have

$$\| u^\dagger - u^0 \|_2 = \| \eta \|_2 = \delta.$$

Thus, we can formulate the above equation as a constraint in the minimization problem

$$\min_u \frac{1}{2} \| \nabla u^0 - Du \|_{L^2(\Omega)}^2 \quad s.t. \quad \| u - u^0 \|_2 \le \delta. \tag{3.12}$$

In this discrete setting the minimization functional writes as

$$\min_u \frac{1}{2} \sum_{i,j} \left| \nabla u_{i,j}^0(x_1, x_2) - (Du(x_1, x_2))_{i,j} \right|^2 \quad s.t. \quad \| u - u^0 \|_2 \le \delta \tag{3.13}$$

where again $|\cdot|$ is the 2-norm.

As a next step, we want to solve this with a projected steepest descent method (see [BV04] for the algorithm); the search direction is given by the negative of the gradient of the minimization functional with respect to u:

$$-\nabla_u \left(\frac{1}{2} \| \nabla u_0 - Du \|_{L^2(\Omega)}^2 \right) = -D^T \left(Du - \nabla u^0 \right).$$

As a constraint for each iteration step k we have $\|u - u^0\|_2 \leq \delta$. The projection of an iterate u^k is necessary if the iterate is no longer in the norm ball $B = \{u \in \Omega \,|\, \|u - u^0\|_2 \leq \delta\}$. The projection is calculated as follows: First, we define the set for the constraint by B (see above) which is a 2-norm ball.

Then, we have

$$\text{proj}_B(u^k) = u^k - \lambda(u^k - u^0)$$

for some $\lambda \in [0,1]$. The idea is sketched in Figure 3.9.

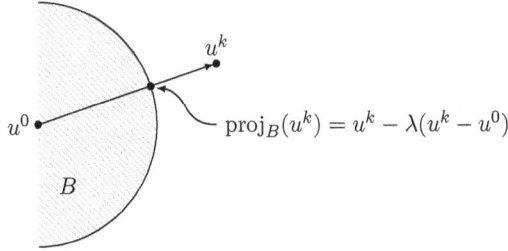

Figure 3.9: Schematic sketch of the projection of an iterate u^k onto the set B.

We calculate $\lambda \in [0,1]$ for which u^k will be on the boundary of B, what is equivalent to $\delta = \|u^0 - \text{proj}_B(u^k)\|_2$. We have

$$\delta = \|u^0 - \text{proj}_B(u^k)\|_2 = \|u^0 - u^k + \lambda(u^k - u^0)\|_2 = \|(u^k - u^0)(1 - \lambda)\|_2$$
$$= (1 - \lambda)\|u^k - u^0\|_2$$
$$\Leftrightarrow \quad \lambda = 1 - \frac{\delta}{\|u^k - u^0\|_2}.$$

The projection then writes as

$$\text{proj}_B(u^k) = \begin{cases} u^k, & \|u^k - u^0\|_2 \leq \delta, \\ u^k - \left(1 - \frac{\delta}{\|u^k - u^0\|_2}\right) \cdot \left(u^k - u^0\right), & \|u^k - u^0\|_2 > \delta, \end{cases}$$
$$= \begin{cases} u^k, & \|u^k - u^0\|_2 \leq \delta, \\ u^0 + \delta \frac{u^k - u^0}{\|u^k - u^0\|_2}, & \|u^k - u^0\|_2 > \delta. \end{cases} \quad (3.14)$$

3.4 Generalization to discontinuous surfaces

In the previous section we derived a data fusion technique to incorporate a given dataset consisting of separately measured surface coordinates and normals. The idea was to calculate directional derivatives according to the neighborhood information given from the structure in the measured dataset (cf. Figure 3.6). The optimization problem writes as

$$\min_u \frac{1}{2} \|Du - \nabla u^0\|_{L^2(\Omega)}^2 \quad s.t. \quad \|u - u^0\|_2 \leq \delta.$$

This produces good results as long as the measured object has no discontinuities. We take a step back and include this property into an updated approach.

The mathematical setup is like before. We consider the measured surface to be a graph over a domain $\Omega \subset \mathbb{R}^2$. To formulate an objective function, we first consider the structure of discontinuous surfaces and what a discontinuity means for each measured point and its neighborhood.

The rough idea is to calculate some kind of *plausible* point coordinates. The plausibility is given with respect to the surface orientation in each point itself and to the reachability of the neighboring points following this orientation, i.e. the surface's slope in this point. If there is a jump in the dataset at a certain measured point the neighboring point on the other side of the jump will be on another level instead of where the surface slope leads to (see Figure 3.10).

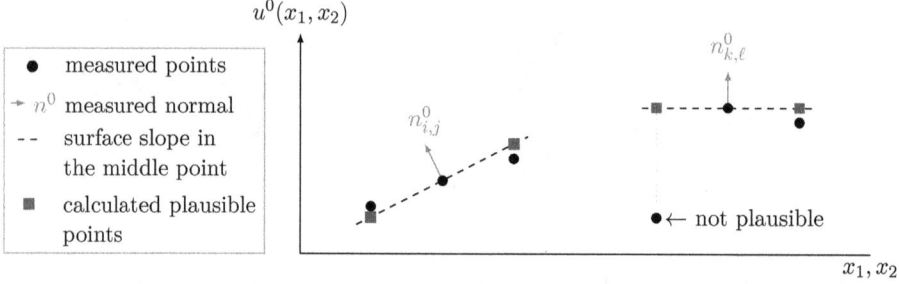

Figure 3.10: Depiction of plausible and implausible points according to surface orientation. We cut through the point cloud and consider just one layer of points to show the concept. A jump is seen in the right scenario where the measured point is on another level according to the function value u^0. The calculated plausible point with respect to the surface orientation given by $n_{k,\ell}^0$ is on the same level as $p_{k,\ell}$.

Let again $u^0 : \Omega \to \mathbb{R}$ be the graph of the measured surface points. Then, we can calculate ∇u^0 as shown in the previous section by a reformulation of the surface normal vectors (see Equation (3.5)). In order to calculate plausible function values u we need to consider all neighbors $p_{k,\ell} \in \mathcal{N}(i,j)$ of a measured point $p_{i,j}$ and the directional vectors $v_{k,\ell}$ which point from the considered point to its neighbor $p_{k,\ell}$ with $(k,\ell) \in I_{i,j}$ (see Equation (3.10)).

Now, we calculate these plausible points by using a first order Taylor expansion with center point $u_{i,j}^0$.

$$u_{k,\ell} = u_{i,j}^0 + \left\langle v_{k,\ell}, \left(\nabla u^0\right)_{i,j} \right\rangle + o \quad \Leftrightarrow \quad u_{k,\ell} - u_{i,j}^0 = \left\langle v_{k,\ell}, \left(\nabla u^0\right)_{i,j} \right\rangle + o,$$

where o is the error term of the Taylor expansion.

The goal is the minimization of the difference between each measured point and its calculated plausible counterpart. This can be done by computing the minimum of

$$\min_u \left\{ \sum_{i,j} \sum_{(k,\ell)\in I_{i,j}} \left| D_{k,\ell}\, u_{i,j} - \left\langle v_{k,\ell}, \left(\nabla u^0\right)_{i,j} \right\rangle - o \right| \quad s.t \quad \|u - u^0\|_2 \le \delta \right\}.$$

We drop the error term and minimize

$$\min_u \left\{ \||Du - V\nabla u^0|\|_1 \quad s.t \quad \|u - u^0\|_2 \le \delta \right\},$$

where D is a sparse matrix in which the considered finite differences of neighboring points are coded. V is a matrix which contains all directional vectors $v_{k,\ell}$. Note that D here is different from the D in Section 3.3. Again, we consider a trusted region in which the points should lay after optimization process, i.e. the same constraint as before.

We switch from the L^2-norm to the L^1-norm in the data term. The reason is the following: Measured sets of point coordinates can have relatively large errors in just one point coordinate within a neighborhood. We call these points outliers. Normally, if there is an error in the measured data these outliers will be within a tolerance value and these points were singular points in a region.

Are these outliers outside a tolerance value it could be a discontinuity within the object and should be detected. In Figure 3.11 we see a neighborhood grid for one $u_{i,j}^0$ and its normal vectors. The level of the points on the right side of the grid should be the same, but the level of the three left points should be below the others. That is the case for surfaces which have a jump at this point. The L^1-norm allows to deal with this discontinuities by preserving the edge since it is insensitive to outliers. The L^2-norm would flatten the areas and round the edges.

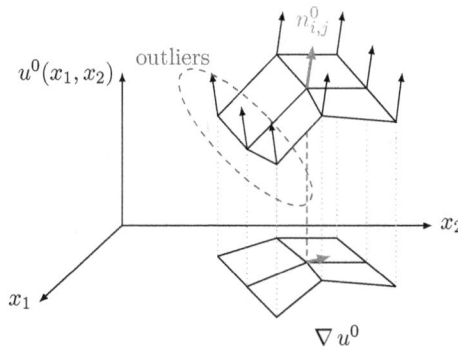

Figure 3.11: Displayed is a neighborhood grid for one $u_{i,j}^0$ and its normal vector $n_{i,j}^0$. In the x_1-x_2-plane lies the gradient ∇u^0 for this cutout as well as the projected normal vector. The three points marked as outliers have a different level, i.e. a different value $u_{k,\ell}^0$ which is significantly smaller than the values of the other points on the right. Hence, the measured object would have a discontinuity at this line.

Now, we have the optimization problem and can move on to solution methods for these. A steepest descent method could be done, but would have to use a slow subgradient descent, since the objective is not differentiable, cf. Problems (3.12) and (3.13). A faster algorithm will be developed in the next section.

3.5 Numerical experiments for data fusion

3.5.1 Preparation

We have proposed approaches to fuse sets of data that are inconsistent according to the measurement accuracy. One approach uses the squared L^2-norm as minimization functional, another approach uses the L^1-norm. For the L^2-norm case we already proposed to use a steepest descent method with a projection according to a set that represents the given constraint. For the L^1-norm we propose to use Chambolle-Pock's method (see Algorithm 1 in Section 2.5 and [CP11]).

One could also think of using other combinations of inner and outer norms. At the moment we consider a functional

$$u \mapsto \||Du - V \nabla u^0|\|_1$$

where $|\cdot|$ is the 2-norm. So, we can write this as

$$u \mapsto \|Du - V \nabla u^0\|_{2,1}.$$

Definition 3.2 (Mixed Norm)
Let (Ω_1, μ_1) and (Ω_2, μ_2) be measure spaces and $u : \Omega_1 \times \Omega_2 \to \mathbb{R}$ measurable with respect to $\mu_1 \otimes \mu_2$. Then, the mixed $L^{p,q}$-norm is defined by

$$\|u\|_{p,q} = \left(\int_{\Omega_2} \left(\int_{\Omega_1} |u(x,y)|^p \, d\mu_1(x) \right)^{q/p} d\mu_2(y) \right)^{1/q}.$$

In order to use Chambolle-Pock's method we need to calculate the Fenchel-Rockafellar conjugate (see Definition 2.43). First, we calculate the dual norm of $\|\cdot\|_{p,q}$ for $p, q > 1$.

Lemma 3.3
Let $p, p', q, q' > 1$, with $1/p + 1/p' = 1$ and $1/q + 1/q' = 1$. Then, the dual norm of $\|\cdot\|_{p,q}$ is $\|\cdot\|_{p',q'}$.

Proof. Let $u, w : \Omega_1 \times \Omega_2 \to \mathbb{R}$ be measurable with respect to $\mu_1 \otimes \mu_2$. In the following, we will use Hölder's inequality (2.26). For any u, w we have

$$\|uw\|_{L^1(\Omega_1)} \leq \|u\|_{L^p(\Omega_1)} \|w\|_{L^{p'}(\Omega_1)},$$

$$\|uw\|_{L^1(\Omega_2)} \leq \|u\|_{L^q(\Omega_2)} \|w\|_{L^{q'}(\Omega_2)}.$$

The definition of the dual norm is (cf. Definition 2.6, Equation (2.2))

$$\|w\|_* = \sup_{\|u\|_{p,q} \leq 1} \langle u, w \rangle_{L^2(\Omega_1 \times \Omega_2)} = \sup_{u \neq 0} \frac{\langle u, w \rangle_{L^2(\Omega_1 \times \Omega_2)}}{\|u\|_{p,q}}.$$

First, we show $\|w\|_* \leq \|w\|_{p',q'}$:

$$\langle u, w \rangle_{L^2(\Omega_1 \times \Omega_2)} \leq \left| \langle u, w \rangle_{L^2(\Omega_1 \times \Omega_2)} \right| = \left| \int_{\Omega_1 \times \Omega_2} u(x,y) w(x,y) \, d(\mu_1 \otimes \mu_2)(x,y) \right|$$

$$\leq \int_{\Omega_1 \times \Omega_2} |u(x,y) w(x,y)| \, d(\mu_1 \otimes \mu_2)(x,y)$$

$$= \int_{\Omega_2} \int_{\Omega_1} |u(x,y) w(x,y)| \, d\mu_1(x) \, d\mu_2(y)$$

$$= \int_{\Omega_2} \|u(\cdot, y) w(\cdot, y)\|_{L^1(\Omega_1)} \, d\mu_2(y)$$

$$\leq \int_{\Omega_2} \underbrace{\|u(\cdot, y)\|_{L^p(\Omega_1)}}_{=:U_p(y)} \underbrace{\|w(\cdot, y)\|_{L^{p'}(\Omega_1)}}_{=:W_{p'}(y)} d\mu_2(y)$$

$$= \int_{\Omega_2} |U_p(y) W_{p'}(y)| \, d\mu_2(y)$$

$$= \|U_p W_{p'}\|_{L^1(\Omega_2)}$$

$$\leq \|U_p\|_{L^q(\Omega_2)} \|W_{p'}\|_{L^{q'}(\Omega_2)}$$

$$= \left(\int_{\Omega_2} |U_p(y)|^q \, d\mu_2(y) \right)^{1/q} \left(\int_{\Omega_2} |W_{p'}(y)|^{q'} \, d\mu_2(y) \right)^{1/q'}$$

$$= \left(\int_{\Omega_2} \|u(\,\cdot\,,y)\|_{L^p(\Omega_1)}^q \right)^{1/q} \left(\int_{\Omega_2} \|w(\,\cdot\,,y)\|_{L^{p'}(\Omega_1)}^{q'} \right)^{1/q'}$$

$$= \left(\int_{\Omega_2} \left(\int_{\Omega_1} |u(x,y)|^p \, d\mu_1(x) \right)^{q/p} d\mu_2(y) \right)^{1/q} \cdots$$

$$\cdots \left(\int_{\Omega_2} \left(\int_{\Omega_1} |w(x,y)|^{p'} \, d\mu_1(x) \right)^{q'/p'} d\mu_2(y) \right)^{1/q'}$$

$$= \|u\|_{p,q} \|w\|_{p',q'}.$$

Hence,

$$\frac{\langle u,\, w \rangle_{L^2(\Omega_1 \times \Omega_2)}}{\|u\|_{p,q}} \leq \|w\|_{p',q'}$$

for $u \neq 0$. This also holds for the supremum of the left-hand side of this inequality and, therefore, $\|w\|_* \leq \|w\|_{p',q'}$. (Note that this inequality holds even for $p,q \in [1,\infty]$ since we only use Hölder's inequality to prove the statement.)

Next, we show that $\langle u,\, w \rangle_{L^2(\Omega_1 \times \Omega_2)} = \|w\|_{p',q'}$ for some $u \neq 0$ with $\|u\|_{p,q} = 1$. Let

$$u(x,y) := \frac{\|w(\,\cdot\,,y)\|_{L^{p'}(\Omega_1)}^{q'-p'}}{\|w\|_{p',q'}^{q'/q}} \operatorname{sign}(w(x,y)) \, |w(x,y)|^{p'-1}.$$

Let us show that the mixed norm of this u is 1:

$$\|u\|_{p,q}^q = \int_{\Omega_2} \left(\int_{\Omega_1} |u(x,y)|^p \, d\mu_1(x) \right)^{q/p} d\mu_2(y)$$

$$= \int_{\Omega_2} \|u(\,\cdot\,,y)\|_{L^p(\Omega_1)}^q \, d\mu_2(y).$$

We calculate the inner norm first. Let

$$\alpha := \frac{1}{\|w\|_{p',q'}^{q'}}.$$

Then,

$$\|u(\,\cdot\,,y)\|_{L^p(\Omega_1)} = \left(\int_{\Omega_1} |u(x,y)|^p \, d\mu_1(x) \right)^{1/p}$$

$$= \left(\int_{\Omega_1} \left| \alpha^{1/q} \, \|w(\,\cdot\,,y)\|_{L^{p'}(\Omega_1)}^{q'-p'} \operatorname{sign}(w(x,y)) \, |w(x,y)|^{p'-1} \right|^p d\mu_1(x) \right)^{1/p}$$

$$= \alpha^{1/q} \, \|w(\,\cdot\,,y)\|_{L^{p'}(\Omega_1)}^{q'-p'} \left(\int_{\Omega_1} |w(x,y)|^{(p'-1)p} \, d\mu_1(x) \right)^{1/p}$$

$$= \alpha^{1/q} \, \|w(\,\cdot\,,y)\|_{L^{p'}(\Omega_1)}^{q'-p'} \underbrace{\left(\int_{\Omega_1} |w(x,y)|^{p'} \, d\mu_1(x) \right)^{1/p}}_{=\|w(\,\cdot\,,y)\|_{L^{p'}(\Omega_1)}^{p'/p}}, \quad \text{since } p' = (p'-1)p,$$

$$= \alpha^{1/q} \, \|w(\,\cdot\,,y)\|_{L^{p'}(\Omega_1)}^{q'-p'+p'/p}$$

$$= \alpha^{1/q} \, \|w(\,\cdot\,,y)\|_{L^{p'}(\Omega_1)}^{q'-1}, \quad \text{since } \frac{p'}{p} = p'-1.$$

Hence, we get

$$
\begin{aligned}
\|u\|_{p,q}^{q} &= \int_{\Omega_2} \|u(\,\cdot\,,y)\|_{L^p(\Omega_1)}^{q}\ \mathrm{d}\mu_2(y)\\
&= \int_{\Omega_2} \left|\alpha^{1/q}\ \|w(\,\cdot\,,y)\|_{L^{p'}(\Omega_1)}^{q'-1}\right|^{q}\ \mathrm{d}\mu_2(y)\\
&= \int_{\Omega_2} \alpha\ \|w(\,\cdot\,,y)\|_{L^{p'}(\Omega_1)}^{(q'-1)q}\ \mathrm{d}\mu_2(y)\\
&= \alpha \underbrace{\int_{\Omega_2} \|w(\,\cdot\,,y)\|_{L^{p'}(\Omega_1)}^{q'}\ \mathrm{d}\mu_2(y)}_{=\|w\|_{p',q'}^{q'}=\alpha^{-1}},\ \text{since}\ \frac{1}{q}+\frac{1}{q'}=1 \Rightarrow q' = (q'-1)q\\
&= 1.
\end{aligned}
$$

We now show that for this u the inner product with w equals $\|w\|_{p',q'}$:

$$
\begin{aligned}
\langle u\,,\,w\rangle_{L^2(\Omega_1\times\Omega_2)} &= \int_{\Omega_2}\int_{\Omega_1} u(x,y)w(x,y)\ \mathrm{d}\mu_1(x)\ \mathrm{d}\mu_2(y)\\
&= \int_{\Omega_2}\int_{\Omega_1} \alpha^{1/q}\ \|w(\,\cdot\,,y)\|_{L^{p'}(\Omega_1)}^{q'-p'}\ \mathrm{sign}(w(x,y))\,|w(x,y)|^{p'-1}\,w(x,y)\ \mathrm{d}\mu_1(x)\ \mathrm{d}\mu_2(y)\\
&= \int_{\Omega_2}\int_{\Omega_1} \alpha^{1/q}\ \|w(\,\cdot\,,y)\|_{L^{p'}(\Omega_1)}^{q'-p'}\ |w(x,y)|^{p'}\ \mathrm{d}\mu_1(x)\ \mathrm{d}\mu_2(y)\\
&= \alpha^{1/q}\int_{\Omega_2} \|w(\,\cdot\,,y)\|_{L^{p'}(\Omega_1)}^{q'-p'}\underbrace{\int_{\Omega_1} |w(x,y)|^{p'}\ \mathrm{d}\mu_1(x)}_{=\|w(\,\cdot\,,y)\|_{L^{p'}(\Omega_1)}^{p'}}\ \mathrm{d}\mu_2(y)\\
&= \alpha^{1/q}\underbrace{\int_{\Omega_2} \|w(\,\cdot\,,y)\|_{L^{p'}(\Omega_1)}^{q'}\ \mathrm{d}\mu_2(y)}_{=\|w\|_{p',q'}^{q'}}\\
&= \|w\|_{p',q'}^{q'(1-1/q)}\\
&= \|w\|_{p',q'},\ \text{since}\ 1-\frac{1}{q}=\frac{1}{q'}.
\end{aligned}
$$

Hence, the dual norm of $\|\cdot\|_{p,q}$ is $\|\cdot\|_{p',q'}$.   □

Since we consider $\|\cdot\|_{2,1}$ we still need to study the case $q=1$.

Lemma 3.4
Let (Ω_1,μ_1) and (Ω_2,μ_2) be measure spaces and $u:\Omega_1\times\Omega_2\to\mathbb{R}$ be measurable with respect to $\mu_1\otimes\mu_2$. Then, the dual norm of $\|u\|_{2,1}$ is

$$
\|w\|_{2,\infty} = \sup_{y\in\Omega_2}\left(\int_{\Omega_1} |w(x,y)|^2\ \mathrm{d}\mu_1(x)\right)^{1/2} = \sup_{y\in\Omega_2} \|w(\,\cdot\,,y)\|_{L^2(\Omega_1)}.
$$

Proof. We know that $\|w\|_* \le \|w\|_{2,\infty}$ (cf. proof of Lemma 3.3). In order to show equality we construct a sequence u^ε such that $\|u^\varepsilon\|_{2,1}\to 1$ and $\langle u^\varepsilon\,,\,w\rangle\to\|w\|_{2,\infty}$. Let

$$
u^\varepsilon(x,y) := \frac{1}{|M_\varepsilon|}\chi_{M_\varepsilon}(y)\ \frac{w(x,y)\,\mathrm{sign}(w(x,y))}{\|w\|_{2,\infty}}
$$

with a measurable set $M_\varepsilon \subset \Omega_2$, $|M_\varepsilon| > 0$ such that for $y \in M_\varepsilon$ it holds that

$$\left| \|w(\cdot, y)\|_{L^2(\Omega_1)} - \operatorname*{ess\,sup}_{y \in M_\varepsilon} \|w(\cdot, y)\|_{L^2(\Omega_1)} \right| < \varepsilon.$$

Now, we show that this u has norm 1 up to ε:

$$\left| \|u^\varepsilon\|_{2,1} - 1 \right| = \left| \int_{\Omega_2} \left(\int_{\Omega_1} |u^\varepsilon(x, y)|^2 \ \mathrm{d}\mu_1(x) \right)^{1/2} \mathrm{d}\mu_2(y) - 1 \right|$$

$$= \left| \int_{\Omega_2} \left(\int_{\Omega_1} \left| \frac{1}{|M_\varepsilon|} \chi_{M_\varepsilon}(y) \frac{w(x, y) \operatorname{sign}(w(x, y))}{\|w\|_{2,\infty}} \right|^2 \mathrm{d}\mu_1(x) \right)^{1/2} \mathrm{d}\mu_2(y) - 1 \right|$$

$$= \frac{1}{\|w\|_{2,\infty}} \left| \frac{1}{|M_\varepsilon|} \int_{M_\varepsilon} \underbrace{\left(\int_{\Omega_1} |w(x, y)|^2 \ \mathrm{d}\mu_1(x) \right)^{1/2}}_{= \|w(\cdot, y)\|_{L^2(\Omega_1)}} \mathrm{d}\mu_2(y) - \|w\|_{2,\infty} \right|$$

$$= \frac{1}{\|w\|_{2,\infty}} \left| \frac{1}{|M_\varepsilon|} \int_{M_\varepsilon} \|w(\cdot, y)\|_{L^2(\Omega_1)} \ \mathrm{d}\mu_2(y) \ldots \right.$$

$$\left. \ldots - \frac{1}{|M_\varepsilon|} \int_{M_\varepsilon} \operatorname*{ess\,sup}_{y \in M_\varepsilon} \|w(\cdot, y)\|_{L^2(\Omega_1)} \ \mathrm{d}\mu_2(y) \right|$$

$$\leq \frac{1}{\|w\|_{2,\infty}} \frac{1}{|M_\varepsilon|} \int_{M_\varepsilon} \underbrace{\left| \|w(\cdot, y)\|_{L^2(\Omega_1)} - \operatorname*{ess\,sup}_{y \in M_\varepsilon} \|w(\cdot, y)\|_{L^2(\Omega_1)} \right|}_{< \varepsilon} \ \mathrm{d}\mu_2(y)$$

$$< \frac{\varepsilon}{\|w\|_{2,\infty}}.$$

Further, we show the equality of $\langle u, w \rangle_{L^2(\Omega_1 \times \Omega_2)} = \|w\|_{2,\infty}$ up to ε:

$$\left| \langle u, w \rangle_{L^2(\Omega_1 \times \Omega_2)} - \|w\|_{2,\infty} \right| = \left| \int_{\Omega_2} \left(\int_{\Omega_1} u^\varepsilon(x, y) w(x, y) \ \mathrm{d}\mu_1(x) \right) \mathrm{d}\mu_2(y) - \|w\|_{2,\infty} \right|$$

$$= \left| \int_{\Omega_2} \left(\int_{\Omega_1} \frac{1}{|M_\varepsilon|} \chi_{M_\varepsilon}(\cdot, y) \frac{w(x, y) \operatorname{sign}(w(x, y))}{\|w\|_{2,\infty}} w(x, y) \ \mathrm{d}\mu_1(x) \right) \mathrm{d}\mu_2(y) - \|w\|_{2,\infty} \right|$$

$$= \frac{1}{\|w\|_{2,\infty}} \left| \frac{1}{|M_\varepsilon|} \int_{M_\varepsilon} \underbrace{\left(\int_{\Omega_1} |w(x, y)|^2 \ \mathrm{d}\mu_1(x) \right)}_{-\|w(\cdot, y)\|_2^2} \mathrm{d}\mu_2(y) - \|w\|_{2,\infty}^2 \right|$$

$$= \frac{1}{\|w\|_{2,\infty}} \left| \frac{1}{|M_\varepsilon|} \int_{M_\varepsilon} \|w(\cdot, y)\|_{L^2(\Omega_1)}^2 \ \mathrm{d}\mu_2(y) - \frac{1}{|M_\varepsilon|} \int_{M_\varepsilon} \operatorname*{ess\,sup}_{y \in M_\varepsilon} \|w(\cdot, y)\|_{L^2(\Omega_1)}^2 \ \mathrm{d}\mu_2(y) \right|$$

$$= \frac{1}{\|w\|_{2,\infty}} \frac{1}{|M_\varepsilon|} \int_{M_\varepsilon} \left| \underbrace{\|w(\cdot, y)\|_{L^2(\Omega_1)}^2}_{=:a} - \underbrace{\operatorname*{ess\,sup}_{y \in M_\varepsilon} \|w(\cdot, y)\|_{L^2(\Omega_1)}^2}_{=:b} \right| \ \mathrm{d}\mu_2(y)$$

$$\leq \frac{1}{\|w\|_{2,\infty}} \frac{1}{|M_\varepsilon|} \int_{M_\varepsilon} |a - b| \ \mathrm{d}\mu_2(y).$$

We know that

$$\left| \|w(\cdot, y)\|_{L^2(\Omega_1)} - \operatorname*{ess\,sup}_{y \in M_\varepsilon} \|w(\cdot, y)\|_{L^2(\Omega_1)} \right| = \left| a^{1/2} - b^{1/2} \right| < \varepsilon.$$

By rationalizing the numerator we obtain $|a - b| < \varepsilon \, |a^{1/2} + b^{1/2}|$. Therefore,

$$
\begin{aligned}
\left| \langle u, w \rangle_{L^2(\Omega_1 \times \Omega_2)} - \|w\|_{2,\infty} \right| &\leq \frac{1}{\|w\|_{2,\infty}} \frac{1}{|M_\varepsilon|} \int_{M_\varepsilon} |a - b| \; \mathrm{d}\mu_2(y) \\
&< \frac{1}{\|w\|_{2,\infty}} \frac{1}{|M_\varepsilon|} \int_{M_\varepsilon} \varepsilon \, \left| a^{1/2} + b^{1/2} \right| \; \mathrm{d}\mu_2(y) \\
&\leq \frac{\varepsilon}{\|w\|_{2,\infty}} \left(\frac{1}{|M_\varepsilon|} \int_{M_\varepsilon} a^{1/2} \; \mathrm{d}\mu_2(y) + \frac{1}{|M_\varepsilon|} \int_{M_\varepsilon} b^{1/2} \; \mathrm{d}\mu_2(y) \right) \\
&\leq \frac{\varepsilon}{\|w\|_{2,\infty}} \left(\frac{1}{|M_\varepsilon|} \int_{M_\varepsilon} a^{1/2} \; \mathrm{d}\mu_2(y) + \|w\|_{2,\infty} \right).
\end{aligned}
$$

Looking closer at the first integral we obtain the following estimate:

$$
\left| a^{1/2} - b^{1/2} \right| < \varepsilon \quad \Rightarrow \quad a^{1/2} < \varepsilon + b^{1/2}
$$

Hence, in the integral

$$
\begin{aligned}
\frac{1}{|M_\varepsilon|} \int_{M_\varepsilon} a^{1/2} \; \mathrm{d}\mu_2(y) &< \frac{1}{|M_\varepsilon|} \int_{M_\varepsilon} \varepsilon + b^{1/2} \; \mathrm{d}\mu_2(y) \\
&\leq \frac{1}{|M_\varepsilon|} \int_{M_\varepsilon} \varepsilon \; \mathrm{d}\mu_2(y) + \int_{M_\varepsilon} b^{1/2} \; \mathrm{d}\mu_2(y) \\
&= \varepsilon + \|w\|_{2,\infty}.
\end{aligned}
$$

Together, we get

$$
\begin{aligned}
\left| \langle u, w \rangle_{L^2(\Omega_1 \times \Omega_2)} - \|w\|_{2,\infty} \right| &\leq \frac{\varepsilon}{\|w\|_{2,\infty}} \left(\frac{1}{|M_\varepsilon|} \int_{M_\varepsilon} \left| a^{1/2} \right| \; \mathrm{d}\mu_2(y) + \|w\|_{2,\infty} \right) \\
&< \frac{\varepsilon}{\|w\|_{2,\infty}} \left(\varepsilon + 2\|w\|_{2,\infty} \right).
\end{aligned}
$$

Hence, the dual norm of $\|\cdot\|_{2,1}$ is $\|\cdot\|_{2,\infty}$.   □

Next, we see that the proposed optimization problem

$$
\min_u \; \|Du - V \nabla u^0\|_1 \quad \text{s.t.} \quad \|u - u^0\|_2 \leq \delta
$$

is of form $\min_x F(x) + G(Kx)$ with

$$
F(u) = \mathcal{I}_{\|\cdot - u^0\|_2 \leq \delta}(u), \quad G(\phi) = \|\|\phi - V \nabla u^0\|\|_1, \quad \text{and} \quad K = D.
$$

In order to use the primal-dual updates within Chambolle-Pock's method we calculate the dual functionals, adjoint operator, and proximal mappings, see Examples 2.44 and 2.48. We have

$$
\begin{aligned}
F^*(s) &= \sup_u \; \langle s, u \rangle - F(u) \\
&= \sup_u \; \langle s, u \rangle - \mathcal{I}_{\|\cdot - u^0\|_2 \leq \delta}(u) \\
&= \sup_{\|t\|_2 \leq \delta} \; \langle s, t \rangle + \langle s, u^0 \rangle \\
&= \delta \|s\|_2 + \langle s, u^0 \rangle
\end{aligned}
$$

and

$$G^*(p) = \sup_{\phi} \langle p, \phi \rangle - G(\phi)$$

$$= \sup_{\phi} \langle p, \phi \rangle - |||\phi - V \nabla u^0|||_1$$

$$= \sup_{\psi} \langle p, \psi + V \nabla u^0 \rangle - |||\psi|||_1$$

$$= \sup_{\psi} \langle p, \psi \rangle + \langle p, V \nabla u^0 \rangle - |||\psi|||_1$$

$$= \mathcal{I}_{|||\cdot|||_1 \le 1}(p) + \langle p, V \nabla u^0 \rangle$$

as well as $K^* = D^T$.

The proximal mappings are

$$\mathrm{prox}_{\tau F}(u) = \mathrm{proj}_{\|\cdot - u^0\|_2 \le \delta}(u) \quad \text{and}$$

$$\mathrm{prox}_{\sigma G^*}(p) = \arg\min_w \frac{1}{2}\|p - w\|_2^2 + \sigma G^*(w)$$

$$= \arg\min_{\|w\|_{2,\infty} \le 1} \frac{1}{2}\|p - w\|_2^2 + \sigma \langle w, V \nabla u^0 \rangle$$

$$= \arg\min_{\|w\|_{2,\infty} \le 1} \frac{1}{2}\|p - w\|_2^2 + \sigma \langle w - p + \frac{1}{2}\sigma V \nabla u^0, V \nabla u^0 \rangle$$

$$= \arg\min_{\|w\|_{2,\infty} \le 1} \frac{1}{2}\|p - w\|_2^2 + \frac{1}{2}\|\sigma V \nabla u^0\|_2^2 - \langle p - w, \sigma V \nabla u^0 \rangle$$

$$= \arg\min_{\|w\|_{2,\infty} \le 1} \frac{1}{2}\left\|(p - w) - \sigma V \nabla u^0\right\|_2^2$$

$$= \arg\min_{\|w\|_{2,\infty} \le 1} \frac{1}{2}\left\|\left(p - \sigma V \nabla u^0\right) - w\right\|_2^2$$

$$= \mathrm{proj}_{\|\cdot\|_{2,\infty} \le 1}\left(p - \sigma V \nabla u^0\right).$$

Here, τ and σ are step sizes. The first projection, onto the norm ball $\|u - u^0\|_2 \le \delta$, was calculated already, see Equation (3.14). The second projection is onto the set $\{w \in \Omega \mid \|w\|_\infty \le 1\}$. The projection of $v = (v_i)_{i=1,\ldots,\#P}$ is given coordinate-wise as

$$\left(\mathrm{proj}_{\|\cdot\|_\infty \le 1}(v)\right)_i = \begin{cases} v_i, & |v_i| \le 1, \\ 1, & v_i > 1, \\ -1, & v_i < -1. \end{cases}$$

The updates for the algorithm are

$$u^{k+1} = \mathrm{prox}_{\tau\Gamma}\left(u^k - \tau D^T p^k\right)$$

$$= \mathrm{proj}_{\|\cdot - u^0\|_2}\left(u^k - \tau D^T p^k\right),$$

$$p^{k+1} = \mathrm{prox}_{\sigma G^*}\left(p^k + \sigma D(2u^{k+1} - u^k)\right) \tag{3.15}$$

$$= \mathrm{proj}_{\|\cdot\|_{2,\infty}}\left(p^k + \sigma D(2u^{k+1} - u^k)\right).$$

3.5.2 Experiments

In this section we will evaluate the results of the previously introduced methods applied to real data. We have two sets of real measured data, first, a plane reference mirror and second, a spherical reference mirror. Additionally, we created a dataset from the plane mirror case which has a discontinuity.

The datasets are given by a struct including a tensor P of type $M \times N \times 3$, i.e. the set of point-coordinates and a tensor N of the same type. For an index (i, j) where $i \in \{1, ..., M\}$ and $j \in \{1, ..., N\}$, the entry $P_{i,j}$ is a triple consisting of x, y and z coordinates. We will call this triple the point-coordinate and we labeled such triples with $p_{i,j}$ in Equation (3.2). Similar, N is a triple with x, y and z coordinates and such triples are the normals. With this structure the neighborhood is given by indices differing by at most one in each direction from the center index (i, j).

Further, not every $p_{i,j}$ or $N_{i,j}$ has a valid entry. There are index pairs (i, j) that have only "Not a Number" (NaN) in all there entries of $p_{i,j}$ or $N_{i,j}$. This happens, for example, when the index pair is at the boundary of the measured object. If there is no measured point there is no value at this matrix index. The same can also happen somewhere else since the measured object can have scratches or holes, or some points are not measured at all because of e.g. measurement errors.

In practice we deal with this issue by removing all indices with NaN values. As first mentioned in Section 3.3, we then talk about the set of valid measured points. To create the two different differential matrices D fromSections 3.3 and 3.4 we remove all rows that correspond with the NaN index pairs in the measured tensor P and N.

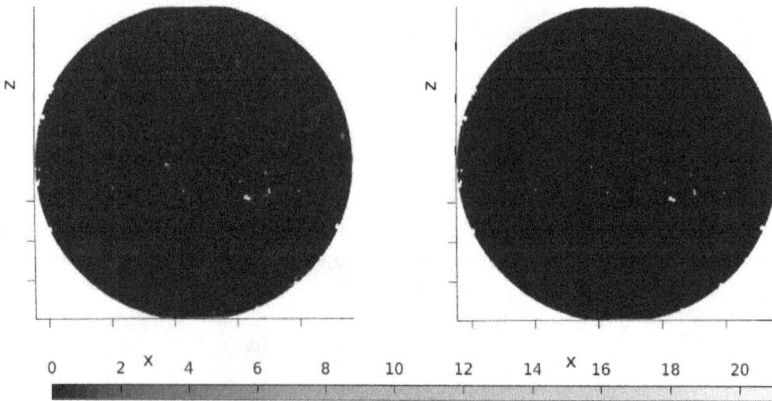

Figure 3.12: Surface plot of the plane reference mirror. The coloring encodes the normal vector error between the measured normal and the normal vectors calculated from the point-coordinates. The color bar is given in degree. Left: The surface before data fusion.

Right: The surface after data fusion with a least-squares optimization functional and projected steepest descent method (see Section 3.3).

Figures 3.12 and 3.13 show the two different measured datasets as well as the results after the data fusion. At first, let us consider the first proposed method (projected steepest descent, see Section 3.3; these results are also in [KLF+14]).

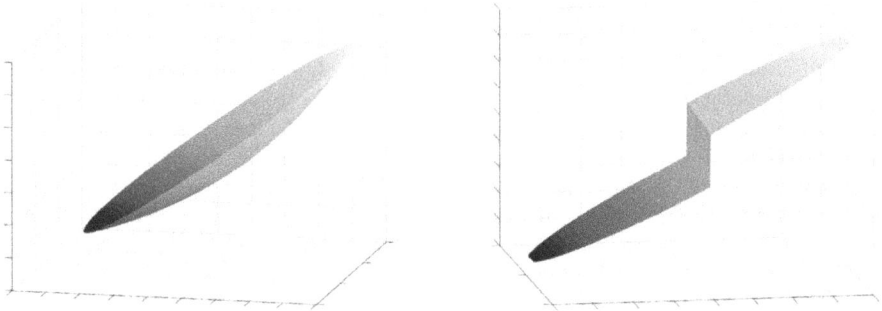

(a) Spherical reference mirror measured data of point coordinates.

(b) Discontinuous data of point coordinates.

Figure 3.13: Surface plots of the measured spherical reference mirror and the point-coordinated with a discontinuity created from the measured plane mirror set.

In the case of the measured plane reference mirror (Figure 3.12) we can see that there are areas where no point-coordinates are available (white areas, especially on the boundary and a relatively large area slightly down to the right from the center of the cyclic object). We had the measured normals and the normals calculated from the measured point-coordinates and used the difference in degree between them to color-code the surface plot. The left plot gives the color-coding before the optimization process. We can see areas with a discrepancy of almost 20°. We also show a close up of this in Figure 3.14 on the left. There also is an area with a discrepancy of 60° between the normals. In both figures the right side is the surface plot of

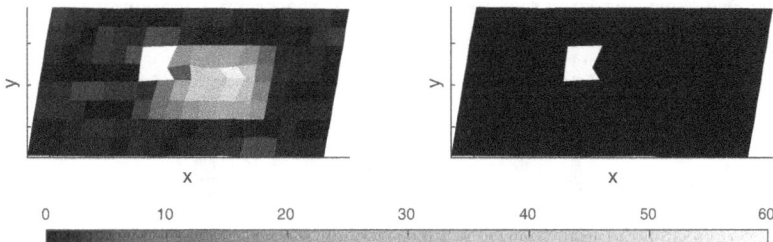

Figure 3.14: Least squares optimized; coloring codes error between normal vectors in degree; one can see that there is a part of the measurement data missing, this could be because of a hole in the object or due to measurement errors.

the plane reference mirror after the optimization process with the least-squares approach. We see that the optimization minimized the discrepancy between the measured normals and the normals calculated from the new point-coordinates.

The same can also be seen in Figure 3.15 where beside the surface plot we also show the normal vectors for one point-coordinate. This selected detail of the whole surface plot is the same as in Figure 3.14. The vector colored in red on the left side of the figure is the measured normal vector. We see that this vector points roughly in an orthogonal direction with respect to the overall surface. The green vector is the normal vector calculated from the point-coordinates in this area. It points in another direction and, hence, the color-coding in the surface plot around this area shows a discrepancy up to almost 60°. On the right we see the detailed view after the optimization process. We already saw that the color-coding represents the minimized discrepancy between the normals. In this figure we also displayed the normal vectors. The green calculated normal vector now lies on top of the measured normal vector in red. The green color is not on the whole line since we wanted to show that the new calculated vector lies on top of the measured vector colored in red. As an a posteriori validation of the assumption

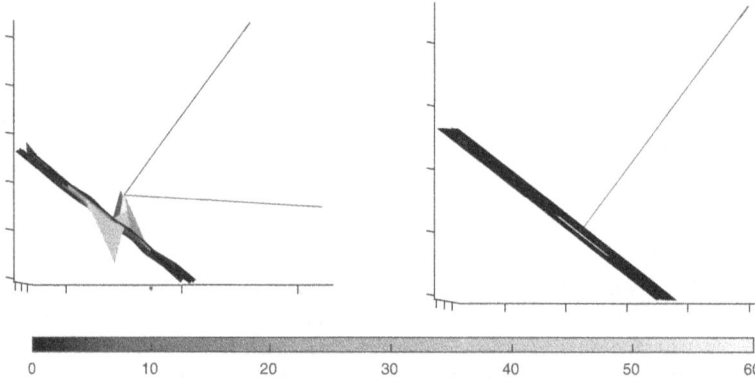

Figure 3.15: Least squared optimized; coloring as before. The red vector is the measured normal vector, the green one the calculated one according to the surrounding point-coordinates. Left before and right after optimization.

that the noise in the third coordinate of the coordinate measurement was indeed Gaussian distributed, we show histograms of the residuum of the original measured third coordinate with the new $u(x_1, x_2)$ in Figure 3.16.

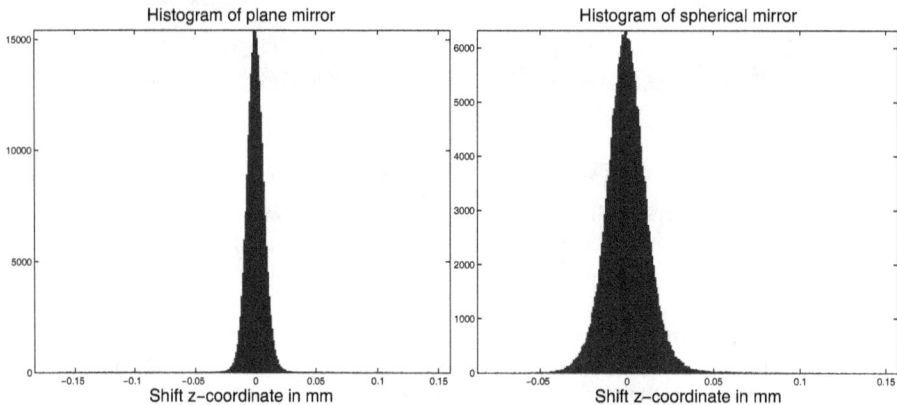

(a) Residuum of the measured z coordinate and the z coordinate after data fusion in the plane reference mirror case.

(b) Residuum of the measured z coordinate and the z coordinate after data fusion in the spherical reference mirror case.

Figure 3.16: Residuum of measured third coordinate after data fusion.

It turns out that the histograms are approximately Gaussian since our data fusion model in fact produces consistent data of point-coordinates and normals. This verifies a posteriori that the assumption of a Gaussian distribution was valid.

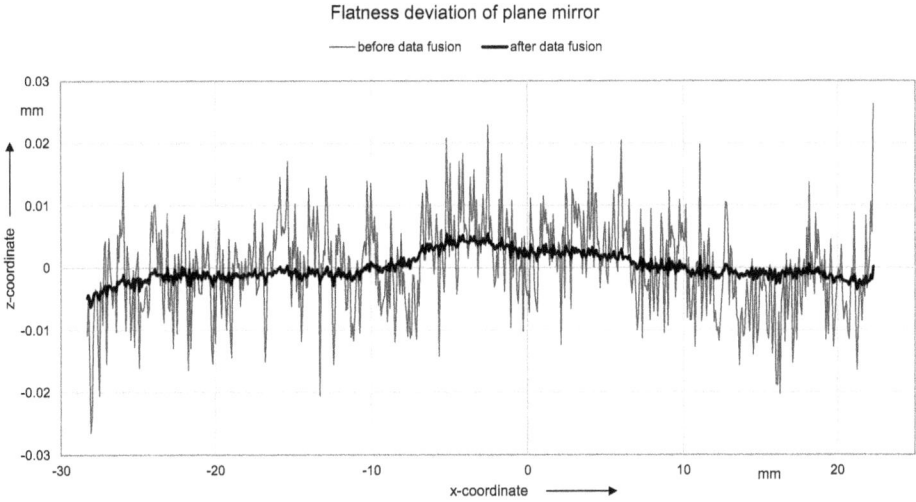

Figure 3.17: Flatness deviation of one row of the plane mirror before and after data fusion.

In Figure 3.17 the deviation of a cross section through the plane mirror is shown. The original data obtained from triangulation shows a noise amplitude of about $20\mu m$. The data fusion process reduces the noise level by approximately one order of magnitude. For the spherical mirror very similar results are obtained.

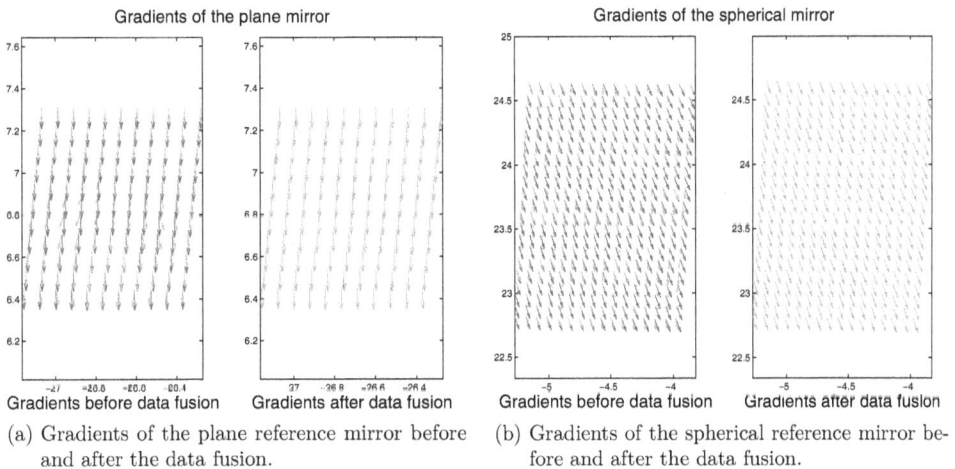

(a) Gradients of the plane reference mirror before and after the data fusion.

(b) Gradients of the spherical reference mirror before and after the data fusion.

Figure 3.18: Gradient vectors before and after data fusion. Measured gradient vectors are displayed in red, calculated gradients with respect to the point coordinates are in green.

In Figure 3.18 the gradients of the measured plane and measured sphere reference mirror are plotted. The two plots in Subfigure 3.18(a) show gradients of the plane mirror before and after the data fusion process. The two plots in Subfigure 3.18(b) show the same for the spherical reference mirror. In every plot the gradients with respect to the measured normals are given in red and the calculated gradients with respect to the point-coordinates are given in green. As one can see, the measured gradients are all pointing in almost the same direction. That is reasonable because the measured objects did not have any visible jumps on the surface. The calculated gradients before the data fusion however differ considerably from the calculated ones. After the data fusion the gradients with respect to the fitted point-coordinates were calculated. On each of the right plots in the subfigures we can see that the newly calculated gradients lie on the measured gradients. Hence, the data fusion process leads to the same information about the orientation of the measured surface. In Figure 3.19 the logarithm of the objective function value for both measured objects is shown. In both semilogarithmic plots we see that the value is still decreasing even after 1000 iterations which means that with enough iterations a very good result will be obtained.

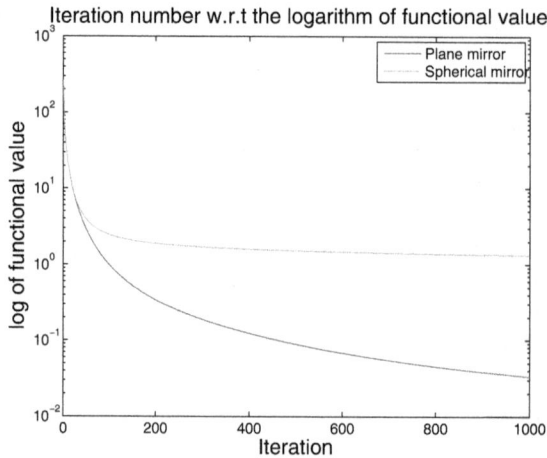

Figure 3.19: Logarithm of the function value during the iteration for a measured plane reference mirror (lower curve, red) and for the spherical reference curve (upper curve, green).

Now, we come to the second proposed data fusion method (see Section 3.4). Again, we tested the plane and spherical reference mirror as well as the discontinuous dataset produced from the plane dataset. We tested two different objective functionals,

$$\min_{u} \frac{1}{2} \left\| \|Du - V \nabla u^0\| \right\|_2^2 \quad s.t. \quad \|u - u^0\|_2 \leq \delta \tag{3.16}$$

$$\text{and} \quad \min_{u} \left\| \|Du - V \nabla u^0\| \right\|_1 \quad s.t. \quad \|u - u^0\|_2 \leq \delta. \tag{3.17}$$

We refer to the first equation as the L^2-functional and to the second as the L^1-functional. For the plane and for the spherical reference mirror the data fusion results are similar to those from the previous approach, and the results for measured objects without jump do not differ much considering both optimization problems (see Figures 3.23 and 3.24). Therefore, we will look more closely at the results of the discontinuous dataset.

In Figure 3.20 the surface after the data fusion process is plotted. Subfigure 3.20(a) shows the result calculated with the L^1-functional. We see that the edge is preserved.

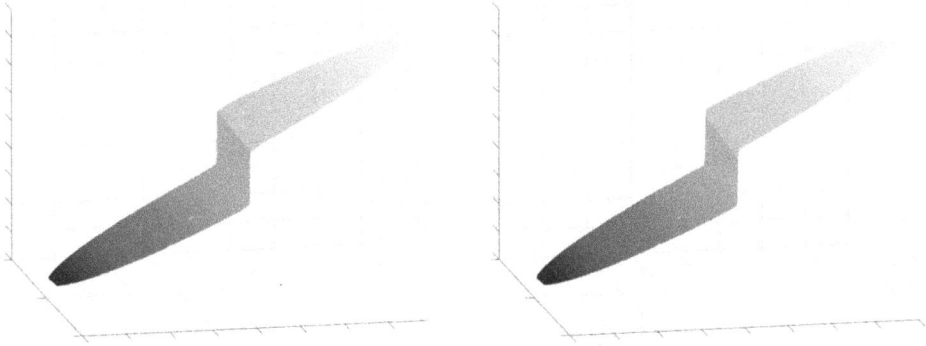

(a) Data fusion with objective functional (3.17) preserves the edge.

(b) Data fusion with objective functional (3.16) rounds the edge.

Figure 3.20: Data fusion results for the discontinuous dataset calculated with the L^1- and the L^2-functional.

In Subfigure 3.20(b) the result calculated with the L^2-functional is shown. Looking closely at the middle segment we see that the edge is distinctively rounded on the top as well as on the bottom part (see the coloring which gives here the value of the third coordinate).

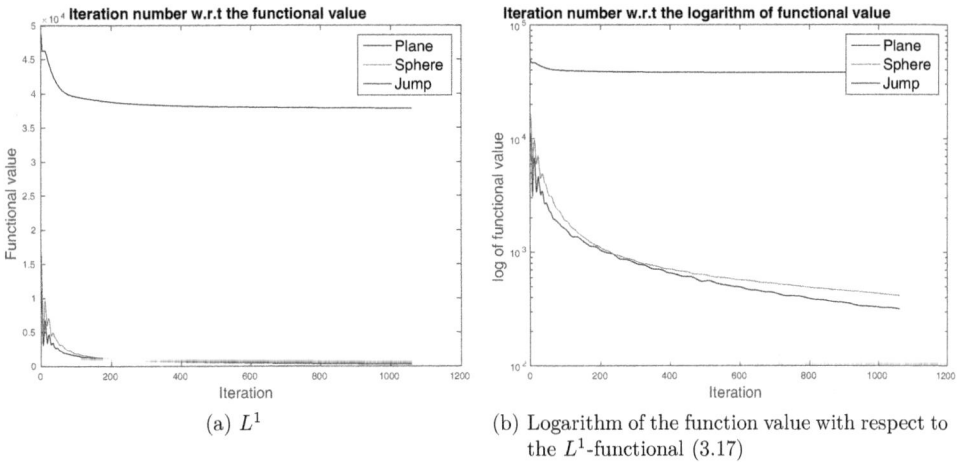

(a) L^1

(b) Logarithm of the function value with respect to the L^1-functional (3.17)

Figure 3.21: Logarithm of the function value during iteration for a measured plane reference mirror (bottom red line), a measured spherical reference mirror (middle green line) and the discontinuous dataset (upper blue line).

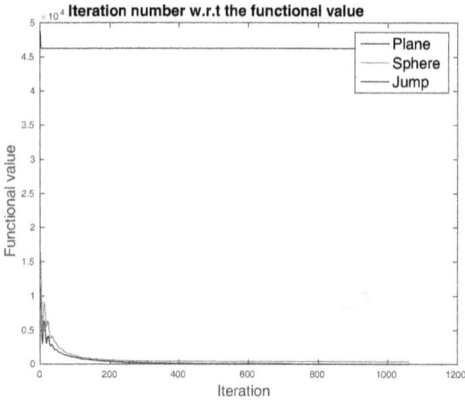

(a) L^2

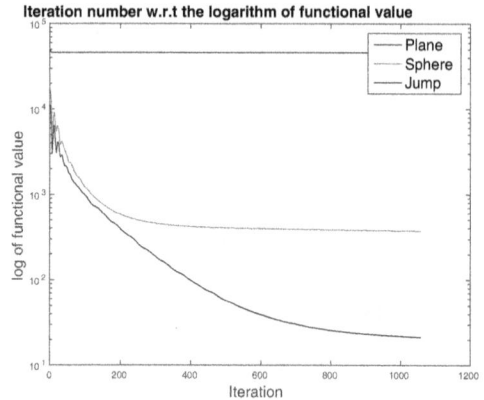

(b) Logarithm of the function value with respect to the L^2-functional (3.16)

Figure 3.22: Logarithm of the function value during iteration for a measured plane reference mirror (bottom red line), a measured spherical reference mirror (middle green line) and the discontinuous dataset (upper blue line).

(a) Surface plot of the plane reference mirror after data fusion with the L^1-functional (3.17).

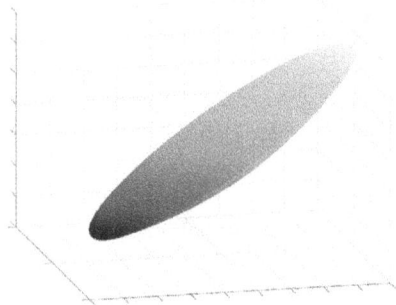

(b) Surface plot of the spherical reference mirror after data fusion with the L^1-functional (3.17).

Figure 3.23: Data fusion results for the plane and the spherical reference mirrors calculated with the L^1-functional.

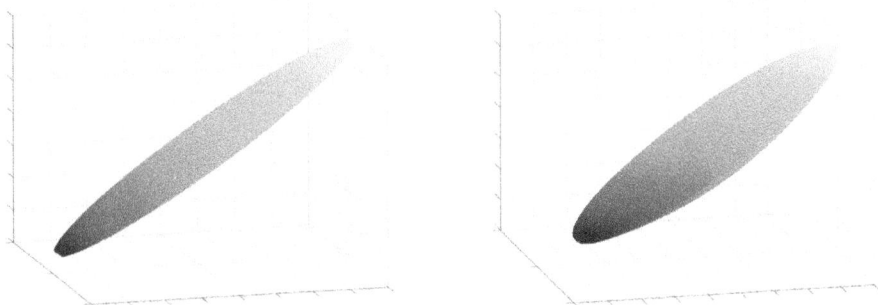

(a) Surface plot of the plane reference mirror after data fusion with the L^2-functional (3.16).

(b) Surface plot of the spherical reference mirror after data fusion with the L^2-functional (3.16).

Figure 3.24: Data fusion results for the plane and the spherical reference mirrors calculated with the L^2-functional.

Two-stage image denoising

Since the introduction in 1992, the Rudin-Osher-Fatemi (ROF) model [ROF92], also known as *total variation denoising*, has found numerous applications. There are different ways to put this model, one is to use the total variation of the image as a regularizer for an image denoising optimization problem. In general it is

$$\min_x F(x) + G(Kx).$$

With u_0 as the input image, defined on a domain Ω we get

$$\min_u \lambda \int_\Omega |\nabla u| \, \mathrm{d}x + \frac{1}{2} \int_\Omega |u(x) - u_0(x)|^2 \, \mathrm{d}x = \min_u \lambda \, \|\|\nabla u\|\|_1 + \frac{1}{2} \|u - u_0\|_2^2. \tag{4.1}$$

One problem in the resulting denoised images is the occurring staircasing effect, i.e. the creation of flat areas separated by jumps. One way to overcome this staircasing was proposed by Lysaker et al. in 2004 [LOT04]. The technique they proposed was a denoising of the image in two separate steps. In a first step, a total variation filter was used to smooth the normal vectors of the level sets of a given noisy image and then, as a second step, a surface was fitted to the resulting normal vectors. The method was formulated in a dynamic way, i.e. by solving a certain partial differential equation to steady state.

A similar approach has been taken in the previous chapter in case of the data fusion process. The measurement device does not only produce approximate point coordinates but also approximate surface normals. It turned out that the incorporation of the surface normals results in an effective, but fairly complicated and non-linear problem. Switching from surface normals to image gradients lead to an effective method. In this chapter we follow the idea of introducing additional information, e.g. gradient information, into the ROF-model (4.1) in order to prevent or minimize the staircasing effect.

4.1 Denoising with prior knowledge of the gradient

Consider the image model

$$u_0 = u^\dagger + \eta,$$

where u_0 is the given noisy image, $u^\dagger$ is the noise-free ground truth image, and η is additive Gaussian white noise. In the situation of images there are methods to obtain a reasonable estimate of the amount of noise, i.e. an estimate on

$$\|u^\dagger - u_0\|_2 = \|\eta\|_2 \tag{4.2}$$

is available. For example, one can use the techniques from [LTO12, LTO13] to estimate the noise level of Gaussian white noise quite accurately from a single image. Using this information it seems that

$$\|u - u_0\|_2 \leq \|u^\dagger - u_0\|_2 = \|\eta\|_2 \tag{4.3}$$

is a sensible condition for the denoised image, since one should not look for an image u further away from $u^\dagger$ than u_0.

This motivates to consider a variant of the ROF-model (4.1) where the discrepancy $\|u - u_0\|_2$ is not a penalty within the objective, but taken into account as a constraint. Thus, we can reformulate the total variation problem as

$$\min_u \||\nabla u|\|_1 \quad s.t. \quad \|u - u_0\|_2 \leq \|u^\dagger - u_0\|_2 = \|\eta\|_2. \tag{4.4}$$

By estimating η as Gaussian noise from the given image u_0 one obtains a parameter free denoising method. Next, assume that we have some additional information on the original image $u^\dagger$ available, namely some estimate v of its gradient. The assumption is similar to the incorporation of the normal vectors within the deflectrometric measurement problem form Chapter 3. This estimate could be taken into account as

$$\min_u \||\nabla u - v|\|_1 \quad s.t. \quad \|u - u_0\|_2 \leq \|\eta\|_2. \tag{4.5}$$

It turns out that this information can be quite powerful. The next lemma shows that if we knew the gradient of $u^\dagger$ and the noise level exactly, our model would recover $u^\dagger$ perfectly, even for arbitrary large noise (and also independent of the type of noise).

Lemma 4.1
Assume that $u^\dagger$ and u_0 fulfill $\int_\Omega u^\dagger = \int_\Omega u_0$ and let $v = \nabla u^\dagger$ and $\delta_1 = \|u^\dagger - u_0\|_2$. Then

$$u^\dagger = \arg\min_u \||\nabla u - v|\|_1 \quad s.t. \quad \|u - u_0\|_2 \leq \delta_1, \tag{4.6}$$

i.e. $u^\dagger$ is the unique solution of the denoising problem.

Proof. The set of minimizers is

$$\arg\min_u \||\nabla u - \nabla u^\dagger|\|_1 \quad s.t. \quad \|u - u_0\|_2 \leq \|u^\dagger - u_0\|_2.$$

Clearly, $u^\dagger$ is within this set, since the optimal value is zero and $u^\dagger$ is feasible, because the constraint is trivially fulfilled.

To show that $u^\dagger$ is indeed the unique solution consider any other u that also produces an objective value of zero. This implies $\nabla u = \nabla u^\dagger$, i.e. $u = u^\dagger + c$ for some constant c. Thus, u fulfills the constraint $\|u - u_0\|_2^2 \leq \|u^\dagger - u_0\|_2^2$ if $\|u^\dagger - u_0 + c\|_2^2 \leq \|u^\dagger - u_0\|_2^2$. We expand the left hand side and get (writing $|\Omega|$ for the measure of Ω)

$$\|u^\dagger - u_0\|_2^2 + 2c \int_\Omega (u^\dagger - u_0) + c^2 |\Omega| \leq \|u^\dagger - u_0\|_2^2.$$

Since the middle integral vanishes by assumption, we see $c = 0$. □

The next lemma shows that $v = \nabla u^\dagger$ is also necessary for the perfect reconstruction.

Lemma 4.2
If $\delta_1 = \|u^\dagger - u_0\|_2$ and $u^\dagger$ solves (4.6) then $v = \nabla u^\dagger$.

Proof. Let $u^\dagger \in \arg\min_u \||\nabla u - v|\|_1$ s.t. $\|u - u_0\|_2 \le \|u^\dagger - u_0\|_2$. We denote by $\mathcal{I}_C(x)$ the indicator function of a set C and set $K = \nabla$, $F(u) = \mathcal{I}_{\|\cdot - u_0\|_2 \le \|u^\dagger - u_0\|_2}(u)$, $G(q) = \||q - v|\|_1$ (i.e. the Fenchel conjugate of G is $G^*(\phi) = \int_\Omega v\phi\,dx + \mathcal{I}_{\||\cdot\||_1 \le 1}(\phi)$). The characterization of optimality by the Fenchel-Rockafellar optimality system [ET76, Remark 4.2] shows that (u^*, ϕ^*) is a primal-dual optimal pair if and only if

$$\begin{cases} 0 \in K^*\phi^* + \partial F(u^*), \\ 0 \in -Ku^* + \partial G^*(\phi^*), \end{cases}$$

which amounts to the inclusions

$$\begin{cases} 0 \in K^*\phi^* + \partial F(u^*), \\ 0 \in -Ku^* + \partial G^*(\phi^*), \end{cases}$$

$$\Leftrightarrow \begin{cases} 0 \in K^*\phi^* + \partial \mathcal{I}_{\|\cdot - u_0\|_2 \le \|u^\dagger - u_0\|_2}(u^*), \\ 0 \in -Ku^* + v + \partial \mathcal{I}_{\||\cdot\||_1 \le 1}(\phi^*). \end{cases}$$

Hence, $(u^\dagger, \phi^*)$ is optimal if and only if

$$\begin{cases} 0 \in -\operatorname{div}\phi^* + \partial \mathcal{I}_{\|\cdot - u_0\|_2 \le \|u^\dagger - u_0\|_2}(u^\dagger), \\ 0 \in -\nabla u^\dagger + v + \partial \mathcal{I}_{\||\cdot\||_1 \le 1}(\phi^*). \end{cases}$$

Since $u^\dagger$ is on the boundary of the domain of the indicator function in the first inclusion, the subgradient is the normal cone, which implies

$$\begin{cases} \exists t \ge 0 : 0 = \operatorname{div}\phi^* + t(u^\dagger - u_0), \\ \nabla u^\dagger(x) = v(x), \\ \qquad \text{if } |\phi^*(x)| < 1, \\ \exists s(x) \ge 0 : \nabla u^\dagger(x) - v(x) = s(x)\phi^*(x), \\ \qquad \text{if } |\phi^*(x)| = 1. \end{cases}$$

By a result of Bourgain and Brezis [BB03, Proposition 1] there is an L^∞ solution ϕ of $-\operatorname{div}\phi = u^\dagger - u_0$, i.e. there exists $L > 0$ such that $|\phi| \le L$ a.e. and hence for $\tilde{\phi} = \phi/(L+1)$ it holds that $-\operatorname{div}\tilde{\phi} = \frac{1}{L+1}(u^\dagger - u_0)$ and $|\tilde{\phi}| < 1$ a.e. Hence, $v = \nabla u^\dagger$ a.e. $\qquad\square$

Remark 4.3. If the condition $\int_\Omega u^\dagger = \int_\Omega u_0$ in Lemma 4.1 does not hold but $\int_\Omega (u^\dagger - u_0) = \varepsilon$, then the proof of that lemma still shows that all solutions of (4.6) are of the form $u^\dagger + c$ with $\left| c - \frac{\varepsilon}{|\Omega|} \right| \le \frac{\varepsilon}{|\Omega|}$.

If $v \ne \nabla u^\dagger$ then any solution $\tilde{u}$ of (4.6) will usually be different from $u^\dagger$, although it will fulfill the trivial estimate

$$\|\tilde{u} - u^\dagger\|_2 \le \|\tilde{u} - u_0\|_2 + \|u_0 - u^\dagger\|_2$$
$$\le 2\|u_0 - u^\mathsf{l}\|_2 = \delta_1.$$

However, the following lemma shows that $\tilde{u} \to u^\dagger$ for $v \to \nabla u^\dagger$ (for constant noise level δ_1).

Lemma 4.4

Assume Ω is convex, $u^\dagger$ and u_0 fulfill $\int_\Omega u^\dagger = \int_\Omega u_0$, let v fulfill $\||v - \nabla u^\dagger|\|_1 \le \varepsilon$ and assume that there exists a solution $\tilde{u}$ of (4.6) with $\delta_1 = \|u^\dagger - u_0\|_2$ for which the constraint is active (i.e. $\|\tilde{u} - u_0\|_2 = \delta_1$).

Then there exists another solution $\bar{u}$ of (4.6) with $\delta_1 = \|u^\dagger - u_0\|_2$ that fulfills $\int_\Omega \bar{u} = \int_\Omega u^\dagger$ and, moreover, it holds that

$$\|\bar{u} - u^\dagger\|_2 \leq \operatorname{diam}(\Omega)\varepsilon$$

where $\operatorname{diam}(\Omega)$ *denotes the diameter of* Ω.

Proof. To obtain $\bar{u}$ we consider $\bar{u} = \tilde{u} + c$ for a suitable constant c. The equality $\int_\Omega \bar{u} = \int_\Omega u^\dagger$ is achieved for $c = \int_\Omega (u^\dagger - \tilde{u})/|\Omega|$. Since $\nabla \bar{u} = \nabla \tilde{u}$ holds $\bar{u}$ is optimal for (4.6) as soon as it is feasible. To check feasibility we calculate

$$\|\bar{u} - u_0\|_2^2 = \|\tilde{u} - u_0 + c\|_2^2$$
$$= \|\tilde{u} - u_0\|_2^2 + 2c \int_\Omega (\tilde{u} - u_0) + c^2 |\Omega|.$$

Since

$$\int_\Omega (\tilde{u} - u_0) = \int_\Omega (\tilde{u} - u^\dagger) + \int_\Omega (u^\dagger - u_0)$$
$$= -c|\Omega| + 0$$

we get

$$\|\bar{u} - u_0\|_2^2 = \|\tilde{u} - u_0\|_2^2 - c^2 |\Omega|$$

which shows feasibility of $\bar{u}$.

Now, we use the Poincaré-Wirtinger inequality in L^1 for which the optimal constant is known from [AD04] to be $\operatorname{diam}(\Omega)/2$, i.e. it holds that

$$\|\bar{u} - u^\dagger\|_1 \leq \tfrac{\operatorname{diam}(\Omega)}{2} \||\nabla(\bar{u} - u^\dagger)|\|_1$$
$$\leq \tfrac{\operatorname{diam}(\Omega)}{2} \left(\||\nabla \bar{u} - v|\|_1 + \||v - \nabla u^\dagger|\|_1 \right).$$

By optimality of $\bar{u}$ and feasibility of $u^\dagger$ we get $\||\nabla \bar{u} - v|\|_1 \leq \||\nabla u^\dagger - v|\|_1$ and hence

$$\|\bar{u} - u^\dagger\|_1 \leq \operatorname{diam}(\Omega)\varepsilon.$$

$\square$

4.2 Denoising of image gradients

In the previous section we studied a constrained version of the total variation image denoising problem. We also introduced the idea of inserting additional information of the ground truth image $u^\dagger$ into the method, namely gradient information. We also studied the sensibility of this incorporation and the gain to the reconstruction quality if $\nabla u^\dagger$ was known.

Since it is unrealistic that the true gradient would be known, the next idea is to create a good estimate v of $\nabla u^\dagger$. One way is to denoise the gradient of the input image, i.e. ∇u_0, specifically by a variational method with a smoothness penalty for the gradient and some discrepancy term. Naturally, a norm of the derivative of the gradient can be used. A first candidate could be the Jacobian of the gradient, i.e.

$$J(\nabla u) = \begin{pmatrix} \partial_1(\partial_1 u) & \partial_2(\partial_1 u) \\ \partial_1(\partial_2 u) & \partial_2(\partial_2 u) \end{pmatrix}$$

which amounts to the Hessian of u. Thus, the matrix is symmetric as soon as u is twice continuously differentiable. However, notice, that the Jacobian of an arbitrary vector field is not necessarily symmetric and hence using $\|J(v)\|$ as smoothness penalty seems unnatural. Instead, we could use a symmetrized Jacobian,

$$\mathcal{E}(v) = \begin{pmatrix} \partial_1 v_1 & \frac{1}{2}(\partial_1 v_2 + \partial_2 v_1) \\ \frac{1}{2}(\partial_1 v_2 + \partial_2 v_1) & \partial_2 v_2 \end{pmatrix},$$

where v_1 and v_2 are the components of v. Note, that for twice differentiable u we have

$$\mathcal{E}(\nabla u) = J(\nabla u) = \mathrm{Hess}(u),$$

i.e. in both cases we obtain the Hessian of u. Imitating the TV-seminorm (and also following the idea of total generalized variation), we take $F(v) = \||\mathcal{E}(v)|\|_1$. Similar to the constraint in (4.6) the denoised gradient should not differ more from the true gradient than ∇u_0 does, thus, we consider the minimization problem with a constraint

$$\min_v \||\mathcal{E}(v)|\|_1 \quad s.t. \quad \||\nabla u_0 - v|\|_1 \leq \delta_2. \tag{4.7}$$

Creating an estimate v by solving the minimization problem above and using this as a prior information in (4.5), we are able to formulate the following two-stage image denoising method:

1. Choose $0 < c \leq 1$ and solve

$$\hat{v} \in \arg\min_v \||\mathcal{E}(v)|\|_1 \quad s.t. \quad \||\nabla u_0 - v|\|_1 \leq c \||\nabla u_0|\|_1 = \delta_2. \tag{4.8}$$

2. Denoise u_0 by solving

$$\hat{u} \in \arg\min_u \||\nabla u - \hat{v}|\|_1 \quad s.t. \quad \|u - u_0\|_2 \leq \delta_1 \approx \|\eta\|_2. \tag{4.9}$$

Instead of using the constrained formulation of the first problem, we can also use a penalized formulation. Thus, the gradient denoising problem writes as:

1. Choose $\alpha > 0$ and solve

$$\hat{v} \in \arg\min_v \||\nabla u_0 - v|\|_1 + \alpha \||\mathcal{E}(v)|\|_1. \tag{4.10}$$

2. Solve (4.9) by using $\hat{v}$ from above.

Regarding the form of both first subproblems we call the first problem *Denoised Gradient Total Variation* (DGTV) and the second problem *Denoised Gradient Total Generalized Variation* (DGTGV). In the next Section 5.5, we will discuss the parameter choice of both problems together with the parameter choice of the following methods. In Chapter 6 we will discuss the quality and performance.

VARIANTS OF TOTAL GENERALIZED VARIATION

In the last chapter we formulated two two-stage methods for image denoising. We considered the total variation image denoising problem but with the twist of incorporating prior information of the gradients of the image into the functional. Further, we reformulated the minimization problem from penalized to constrained type, which looked like

$$\min_{u} \||\nabla u - v|\|_1 \quad s.t. \quad \|u - u_0\|_2 \leq \delta_1.$$

It turned out that the information of a good approximation of the true gradient $\nabla u^\dagger$ is a powerful information. Hence, a denoising of ∇u_0 in two different ways was proposed.

In this chapter we want to study different ways of combining the two-stage methods into one optimization problem where the image and the gradient are denoised simultaneously.

5.1 Total generalized variation

One way to overcome the staircasing effect is to introduce higher-order derivatives of u into the optimization problem. So, instead of considering the total variation as a regularizer, cf. minimization problem (4.1), Kristian Bredies et al. introduced the total generalized variation in 2010 [BKP10].

Let us review the definition from Chapter 2.

Definition 5.1 (Total generalized variation [BKP10])
Let $\Omega \subset \mathbb{R}^d$ be a domain, let $k \geq 1$, and let $\alpha_0, \dots \alpha_{k-1} > 0$. Then, the total generalized variation of order k with weight α for $u \in L^1_{\text{loc}}(\Omega)$ is defined as the value of the functional

$$\text{TGV}^k_\alpha(u) = \sup \left\{ \int_\Omega u \operatorname{div}^k v \, dx \ \middle|\ v \in C^k_C(\Omega, \text{Sym}^k(\mathbb{R}^d)), \|\operatorname{div}^l v\|_\infty \leq \alpha_l, l = 0, \dots, k-1 \right\},$$

where the supremum admits the value ∞ where the respective set is unbound from above.

The space

$$\text{BGV}^k_\alpha(\Omega) = \left\{ u \in L^1(\Omega) \ \middle|\ \text{TGV}^k_\alpha(u) < \infty \right\}, \quad \|u\|_{\text{BGV}^k_\alpha} = \|u\|_1 + \text{TGV}^k_\alpha(u) \tag{5.1}$$

is called the space of functions of bounded generalized variation of order k with weight α.

Here, $\text{Sym}^k(\mathbb{R}^d)$ denotes the space of symmetric tensors of order k with arguments in $\mathbb{R}^d$. Further, $\operatorname{div}^l v$ is defined as the l-divergent. Later we only need the special case $k = 2$ which writes with interpretation as symmetric matrix as

$$(\operatorname{div} v)_i = \sum_{j=1}^d \frac{\partial v_{ij}}{\partial x_j}, \quad \operatorname{div}^2 v = \sum_{i=1}^d \frac{\partial^2 v_{ii}}{\partial x_i^2} + \sum_{i<j} 2 \frac{\partial^2 v_{i,j}}{\partial x_i \partial x_j}. \tag{5.2}$$

Additionally, for $k = 0$ it is the identity and for $k = 1$ it is the usual divergence.

Remark 5.2. For $k = 1$ and $\alpha > 0$, we see that

$$\mathrm{TGV}^1_\alpha(u) = \alpha \sup \left\{ \int_\Omega u \operatorname{div} v \, dx \,\middle|\, v \in C^1_C(\Omega, \mathrm{Sym}^1(\mathbb{R}^d)), \|v\|_\infty \le 1 \right\} = \alpha \, \mathrm{TV}(u)$$

For the numerical methods they consider mainly the TGV^2_α regularization term and a quadratic L^2 data fidelity term. The $\mathrm{TGV}^2_\alpha - L^2$ image denoising problem then is

$$\min_{u \in L^2(\Omega)} \frac{1}{2} \|u - u_0\|^2_2 + \mathrm{TGV}^2_\alpha \tag{5.3}$$

for some positive $\alpha = (\alpha_0, \alpha_1)$. This problem can be rewritten as

$$\min_{v \in H^2_{\mathrm{div},c}(\Omega, S^{d \times d})} \frac{1}{2} \|\operatorname{div}^2 v - u_0\|^2_2 + \mathcal{I}_{\|v\|_\infty \le \alpha_0}(v) + \mathcal{I}_{\|\operatorname{div} v\|_\infty \le \alpha_1}(v). \tag{5.4}$$

5.2 Types of minimization problems: Tikhonov, Morozov, Ivanov

Consider minimization problem (2.18), i.e.

$$\min_u \Phi(Ku - u_0) + \lambda \Psi(u). \tag{5.5}$$

Given two objective functions Φ, the discrepancy functional, and Ψ, the regularization functional, we have at least three different types of minimization problems (cf [LW13]):

Tihonov regularization [Tik63]. For some $\lambda > 0$ it sets the problem exactly as in Equation (5.5). Thus, we weight between a 'good data fit' and a 'good fit to prior knowledge'. For example, the TGV-denoising problem, Equation (5.9), is a Tikhonov regularization.

Ivanov regularization [Iva62]. Let $\tau > 0$ and consider

$$\min_u \Phi(Ku - u_0) \quad s.t. \quad \Psi(u) \le \tau. \tag{5.6}$$

Thus, we choose the solution which fits the data best, but also fits the prior knowledge up to a predefined amount.

Morozov regularization [Mor67]. Let $\delta > 0$ and consider

$$\min_u \Psi(u) \quad s.t. \quad \Phi(Ku - u_0) \le \delta. \tag{5.7}$$

We choose a solution which fits the prior knowledge best but also explains the data up to a predefined amount.

Considering the three variational schemes one can show that if there exists a unique solution of one of these problems then there exists a unique solution of the other schemes, see Theorem 2.3 in [LW13].

In Chapter 4 we already saw different formulations of denoising minimization problems. In this chapter we will take a closer look to these different formulations.

5.3 Constrained total generalized variation

Considering the total variation problem (4.1) with the insertion of a denoised gradient, i.e. an estimate $\hat{v}$ of $\nabla u^{\dagger}$. As an additional information, we get the following minimization problem:

$$\min_{u} \frac{1}{2} \|u - u_0\|_2 + \lambda \|\nabla u - \hat{v}\| . \tag{5.8}$$

One way to get such an estimate is to solve (4.10), i.e.

$$\hat{v} = \arg\min_{v} \|\|\nabla u_0 - v\|\|_1 + \alpha \|\|\mathcal{E}(v)\|\|_1 .$$

If we combine these two steps, the total variation denoising with additional information and the calculation of said information, we obtain

$$\min_{u,v} \frac{1}{2} \|u - u_0\|_2 + \lambda \|\|\nabla u - v\|\|_1 + \alpha \|\|\mathcal{E}(v)\|\|_1 .$$

Renaming the penalization parameters $\alpha_0 := \alpha$ and $\alpha_1 := \lambda$, we get the total generalized variation image denoising problem (cf. [BKP10, KBPS11]):

$$\min_{u,v} \frac{1}{2} \|u - u_0\|_2 + \alpha_1 \|\|\nabla u - v\|\|_1 + \alpha_0 \|\|\mathcal{E}(v)\|\|_1$$
$$= \min_{u} \frac{1}{2} \|u - u_0\|_2 + \mathrm{TGV}_\alpha(u) \tag{5.9}$$

with the total generalized variation penalty

$$\mathrm{TGV}_\alpha(u) = \min_{v} \alpha_1 \|\|\nabla u - v\|\|_1 + \alpha_0 \|\|\mathcal{E}(v)\|\|_1 . \tag{5.10}$$

Next, consider a purely constrained variant of the image denoising problem. For this, we take the constraint total variation with the insertion of the prior information (4.9) and combine it with the first step (4.8) from DGTV. This leads to

$$\min_{u,v} \|\|\mathcal{E}(v)\|\|_1 \quad s.t. \quad \|u - u_0\|_2 \leq \delta_1$$
$$\|\|\nabla u - v\|\|_1 \leq \delta_2. \tag{5.11}$$

Consequently, we define the *constrained total generalized variation* (CTGV) as

$$\mathrm{CTGV}_\delta(u) = \min_{v} \|\|\mathcal{E}(v)\|\|_1 \quad s.t. \quad \|\|\nabla u - v\|\|_1 \leq \delta \tag{5.12}$$

and problem (5.11) becomes

$$\min_{u} \mathrm{CTGV}_{\delta_2}(u) \quad s.t. \quad \|u - u_0\|_2 \leq \delta_1. \tag{5.13}$$

Let us take a closer look at (5.11): If we reformulate the problem with penalization instead of constraints, we obtain

$$\min_{u,v} \|\|\mathcal{E}(v)\|\|_1 + \frac{\beta_0}{2} \|u - u_0\|_2^2 + \beta_1 \|\|\nabla u - v\|\|_1 .$$

The parameters are $\beta_0 = 1/\alpha_0$ and $\beta_1 = \alpha_1/\alpha_0$ and we get exactly the TGV problem (5.9). According to minimizers of the TGV and CTGV problems, both are equivalent in the sense of minimizers of the TGV and CTGV functionals and in the sense of minimizers of the associated denoising problems as the following lemmas show.

Lemma 5.3

The functionals TGV *and* CTGV *are equivalent in the sense that for any u it holds that for any $\alpha > 0$ there exists $\delta > 0$ such that each minimizer v_α of the* $\mathrm{TGV}_\alpha(u)$ *minimization problem is also a minimizer of the* $\mathrm{CTGV}_\delta(u)$ *minimization problem and vice versa.*

Results that relate solutions for penalized and constrained problems are fairly common, see e.g. [LW13].

Proof. Consider $\alpha_0 = 1$ and $\alpha_2 = \alpha$. Let v_α be a minimizer of the TGV functional (5.10), i.e.

$$\mathrm{TGV}_{1,\alpha}(u) = \alpha \, |||\nabla u - v_\alpha|||_1 + |||\mathcal{E}(v_\alpha)|||\,.$$

Then for $\delta(\alpha) = |||\nabla u - v_\alpha|||_1$ it is clear that v_α is also feasible and optimal for (5.12).
  Conversely, if $\alpha(\delta)$ is the Lagrange multiplier for the constraint in (5.12) and v_δ is optimal there, then v_δ is also optimal for the problem (5.10) with $\alpha_1 = \alpha(\delta)$ and $\alpha_0 = 1$. $\quad\square$

Lemma 5.4

The TGV *denoising problem* (5.9) *and the* CTGV *denoising problem* (5.11) *are equivalent in the sense that for any minimizer u_α of* (5.9) *there exists δ_1, δ_2 such that u_α is a solution of* (5.11). *Conversely, for any δ_1, δ_2 such that the constraints in* (5.11) *are active for a minimizer u_δ there exist α_0, α_1 such that u_δ is a solution of* (5.9).

Proof. This proof is similar to the previous one: Assume that u_α is a solution of (5.9) and v_α is the corresponding optimal vector field. Now, set

$$\delta_1 = \|u_\alpha - u_0\|_2 \qquad \text{and}$$
$$\delta_2 = |||\nabla u_\alpha - v_\alpha|||_1\,.$$

Then, the pair (u_α, v_α) is feasible and optimal for (5.11).
  Conversely, if (u_δ, v_δ) is an optimal pair for (5.11) with active constraints, set α_0 and α_1 to the respective Lagrange multipliers and observe that (u_δ, v_δ) is also optimal for (5.9). $\quad\square$

5.4 Morozov total generalized variation

In the previous section we saw how to combine both steps of the two-stage denoising method DGTV (4.8), (4.9) to one denoising method which has two constraints, CTGV (5.13). This section will introduce another way to combine the functionals into an image denoising method.
  Consider now both steps of the DGTV problem (4.10), (4.9). The first step there was to minimize the term

$$\min_v \, |||\nabla u_0 - v|||_1 + \alpha \, |||\mathcal{E}(v)|||_1 = \mathrm{TGV}^2_{(\alpha,1)}(u_0).$$

Combining this with an update of the image u and the second step of DGTV we define the *Morozov total generalized variation* (MTGV) image denoising problem as

$$\min_{u,v} \, \{|||\nabla u - v|||_1 + \alpha \, |||\mathcal{E}(v)|||_1 \quad s.t. \quad \|u - u_0\|_2 \le \delta_1\}$$
$$= \min_{u,v} \, \left\{\mathrm{TGV}^2_{(\alpha,1)}(u) \quad s.t. \quad \|u - u_0\|_2 \le \delta_1\right\}. \tag{5.14}$$

Corollary 5.5

The TGV *denoising problem* (5.9) *and the* MTGV *denoising problem* (5.14) *are equivalent in the sense that for any minimizer u_α of* (5.9) *there exist α and δ_1 such that u_α is also a solution of* (5.14). *Conversely, for any α, δ_1 such that the constraint in* (5.14) *is active for a minimizer u_{δ_1} there exist α_0 and α_1 such that u_{δ_1} is a solution of* (5.9).

A few words on the proof: Consider the TGV term (5.10) set $(\alpha_0, \alpha_1) = (\alpha, 1)$ to obtain the associated term in the MTGV problem (5.14). Setting $\delta_1 = \|u_\alpha - u_0\|_2$ where u_α is a solution of (5.9), we see that u_α is feasible for (5.14).

How to choose the according parameters in practice will be described in Section 5.5.

5.5 Other possible combinations and parameter choice

Our denoising methods use the following terms:

$$\|u - u_0\|_2, \tag{5.15}$$
$$\||\nabla u - v|\|_1, \tag{5.16}$$
$$\text{and} \quad \||\mathcal{E}(v)|\|_1. \tag{5.17}$$

The first term (5.15) is a discrepancy term. The second term (5.16) is a smoothing prior for the two-step denoising or a gradient fitting term for the other methods introduced in this section. The third term (5.17) is a smoothing term for the gradient estimate. Let us recall the two-stage methods in Chapter 4. There, we used these functionals but in a kind of separate way. As a reminder, the DGTV problem writes as

1. Choose $0 < c \le 1$ and solve

$$\hat{v} \in \arg\min_v \||\mathcal{E}(v)|\|_1 \quad s.t. \quad \||\nabla u_0 - v|\|_1 \le c \||\nabla u_0|\|_1 = \delta_2.$$

2. Denoise u_0 by solving

$$\hat{u} \in \arg\min_u \||\nabla u - \hat{v}|\|_1 \quad s.t. \quad \|u - u_0\|_2 \le \delta_1 \approx \|\eta\|_2. \tag{4.9}$$

And the DGTGV problem writes as

1. Choose $\alpha > 0$ and solve

$$\hat{v} \in \arg\min_v \||\nabla u_0 - v|\|_1 + \alpha \||\mathcal{E}(v)|\|_1.$$

2. Solve (4.9) by using $\hat{v}$ from above.

Combining the functionals above into one minimization problem gives us seven possibilities: We could use every term in the optimization functional as penalizations, one of the three terms as constraint, or two of the three terms as constraints. Hence, we obtain:

$$\min_{u,v} \frac{1}{2}\|u - u_0\|_2^2 + \alpha_1 \||\nabla u - v|\|_1 + \alpha_0 \||\mathcal{E}(v)|\|_1, \tag{5.18}$$

$$\min_{u,v} \frac{1}{2}\|u - u_0\|_2^2 + \gamma \||\nabla u - v|\|_1 \quad s.t. \quad \||\mathcal{E}(v)|\|_1 \le \beta, \tag{5.19}$$

$$\min_{u,v} \frac{1}{2} \|u - u_0\|_2^2 + \gamma \||\mathcal{E}(v)|\|_1 \quad s.t. \quad \||\nabla u - v|\|_1 \leq \delta_2, \qquad (5.20)$$

$$\min_{u,v} \||\nabla u - v|\|_1 + \alpha \||\mathcal{E}(v)|\|_1 \quad s.t. \quad \|u - u_0\|_2 \leq \delta_1, \qquad (5.21)$$

$$\min_{u,v} \frac{1}{2} \|u - u_0\|_2^2 \quad s.t. \quad \||\nabla u - v|\|_1 \leq \delta_2$$
$$\||\mathcal{E}(v)|\|_1 \leq \beta, \qquad (5.22)$$

$$\min_{u,v} \||\nabla u - v|\|_1 \quad s.t. \quad \|u - u_0\|_2 \leq \delta_1$$
$$\||\mathcal{E}(v)|\|_1 \leq \beta, \qquad (5.23)$$

$$\min_{u,v} \||\mathcal{E}(v)|\|_1 \quad s.t. \quad \|u - u_0\|_2 \leq \delta_1$$
$$\||\nabla u - v|\|_1 \leq \delta_2. \qquad (5.24)$$

The other possibilities to combine the three functionals come with penalization and constraint parameters. In order to use these problem formulations it is necessary to choose or determine these parameters.

We have seen the optimization problems (5.18), (5.21), and (5.24) before as they are the TGV, MTGV and CTGV problems, respectively. Additionally, we considered two-stage problem formulations with these functionals in Chapter 4. We did not consider Problems (5.19), (5.22) and (5.23) since they contain the penalty $\||\mathcal{E}(v)|\|_1 \leq \beta$ and we did not find a reasonable way to choose the parameter β.

Let us recall the image model. Given a noisy image u_0 we assume that there is a clean image $u^\dagger$ which is corrupted by additive Gaussian noise η, thus it is given by

$$u_0 = u^\dagger + \eta.$$

Reformulating this equation and considering the norm, we get

$$\|u_0 - u^\dagger\|_2 = \|\eta\|_2.$$

This is one of the terms which we use as a constraint with upper bound δ_1 in Equations (5.21), (5.23) and (5.24).

We consider η to be normal distributed, hence, $\eta_i \sim \mathcal{N}(0, \sigma^2)$ where σ is the standard deviation. We are able to write the norm of noise η as

$$\|\eta\|_2 = \sigma \|\bar{\eta}\|_2$$

where $\bar{\eta}_i \sim \mathcal{N}(0, 1)$. It follows that

$$\|\bar{\eta}\|_2 = \sum_{i=1}^{k} \bar{\eta}_i \sim \chi_k^2,$$

i.e. $\|\bar{\eta}\|_2$ is distributed according to a χ_k^2-distribution. Since $\|\eta\|_2 = \sigma \|\bar{\eta}\|_2$ it follows for the expectation value that

$$\mathbb{E}\left[\|\eta\|_2\right] = \sigma \mathbb{E}\left[\|\bar{\eta}\|_2\right].$$

Thus, we calculate $\mathbb{E}\left[\|\bar{\eta}\|_2\right]$ since we know the probability density function of a χ_k^2-distributed random variable X with k degrees of freedom.

Let $X = \|\bar{\eta}\|_2^2$. Then, for $g(X) = \sqrt{X}$ it holds that

$$\mathbb{E}\left[\sqrt{X}\right] = \mathbb{E}\left[g(X)\right] = \int_{\mathbb{R}} g(x) f(x) \, dx$$

where $f(x)$ is the probability density function of the χ_k^2-distribution. Here, it is

$$f(x; k) = \begin{cases} \dfrac{x^{\frac{k}{2}-1} e^{-\frac{x}{2}}}{2^{\frac{k}{2}} \Gamma\left(\frac{k}{2}\right)}, & x > 0, \\ 0, & x \leq 0, \end{cases}$$

where $\Gamma(\frac{k}{2})$ denotes the gamma function, which has closed-form values for integer k. We calculate

$$\mathbb{E}\left[\sqrt{X}\right] = \mathbb{E}\left[g(X)\right]$$
$$= \int_{\mathbb{R}} \sqrt{x} f(x; k) \, dx$$
$$= \int_0^\infty \sqrt{x} \, \frac{x^{\frac{k}{2}-1} e^{-\frac{x}{2}}}{2^{\frac{k}{2}} \Gamma\left(\frac{k}{2}\right)} \, dx$$
$$= \frac{1}{2^{\frac{k}{2}} \Gamma\left(\frac{k}{2}\right)} \int_0^\infty x^{\frac{k}{2}-\frac{1}{2}} e^{-\frac{x}{2}} \, dx.$$

In general, the gamma function is defined as the integral

$$\Gamma(z) = \int_0^\infty t^{z-1} e^{-t} \, dt.$$

The above integral can be rewritten as a gamma function with the help of the substitution $t = \frac{x}{2}$, $dx = 2 \, dt$:

$$\int_0^\infty x^{\frac{k-1}{2}} e^{-\frac{x}{2}} \, dx = \int_0^\infty (2t)^{\frac{k-1}{2}} e^{-t} \cdot 2 \, dt$$
$$= 2^{\frac{k+1}{2}} \int_0^\infty t^{\frac{k+1}{2}-1} e^{-t} \, dt$$
$$= 2^{\frac{k+1}{2}} \Gamma\left(\frac{k+1}{2}\right).$$

Inserting this, we get

$$\mathbb{E}[\|\bar{\eta}\|_2] = \frac{1}{2^{\frac{k}{2}} \Gamma\left(\frac{k}{2}\right)} 2^{\frac{k+1}{2}} \Gamma\left(\frac{k+1}{2}\right)$$
$$= \sqrt{2} \, \frac{\Gamma\left(\frac{k+1}{2}\right)}{\Gamma\left(\frac{k}{2}\right)}.$$

With Gautschi's inequality [Gau59], where $0 < s < 1$,

$$x^{1-s} < \frac{\Gamma(x+1)}{\Gamma(x+s)} < (x+1)^{1-s},$$

we get

$$(k-1)^{\frac{1}{2}} < \sqrt{2}\,\frac{\Gamma\left(\frac{k+1}{2}\right)}{\Gamma\left(\frac{k}{2}\right)} < (k+1)^{\frac{1}{2}}.$$

It follows that

$$\mathbb{E}\left[\|\bar{\eta}\|_2\right] \approx \sqrt{k},$$

and hence,

$$\mathbb{E}\left[\|\eta\|_2\right] \approx \sigma\sqrt{k}.$$

For a given discrete image there are methods from [LTO12, LTO13] to estimate the standard deviation σ of the noise η. Then, $\|\eta\|_2 \approx \sigma\sqrt{k}$ where again k is the number of pixels.

The parameter δ_2 and the parameter α from the DGTGV problem (4.10) are set via experimental results. We calculated the parameter c, where $\delta_2 = c\,\||\nabla u_0|\|_1$ is the constraint parameter, such that the distance of the denoised image to the ground truth noise-free image is as small as possible. In that sense the parameter c is optimal. We calculated c for various images of different size and different noise levels, see Figure 5.1. As expected, all $c \geq 1$ lead to

DGTV, optimal c

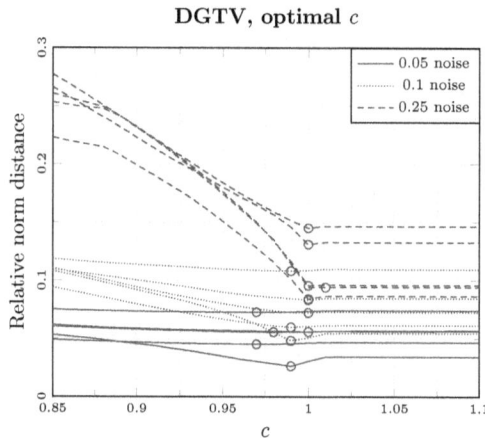

Figure 5.1: Optimal values for problem parameter c in (4.8) for various images and different noise level for the DGTV method.

the same result. In this case $v = 0$ is a feasible solution and optimal. Therefore, the two-stage method becomes pure total variation denoising. If we choose $c < 1$, we transfer a bit of structure of the input image into the gradient as an additional information for the image denoising step. Hence, $c \approx 0.99$ seems like a sensible choice.

A similar experiment for the parameter α in the DGTGV method revealed that all these optimal values are close to 1, see Figure 5.2. For $\alpha > 1$ the change in the norm distance is minimal. Hence, the denoised image will be similar to the denoised image with $\alpha = 1$. Smaller values $\alpha < 1$ lead to worse reconstructions. Therefore, we use a default value of $\alpha = 1$.

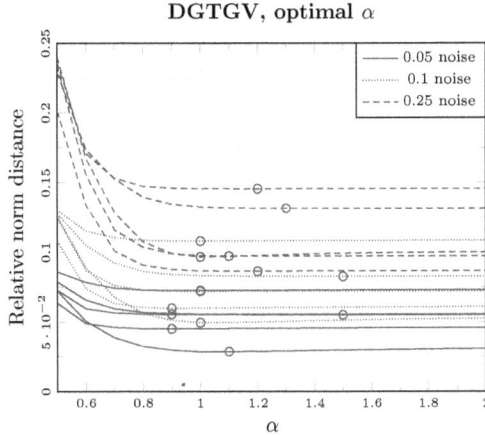

Figure 5.2: Optimal values for problem parameter α in (4.10) for various images and different noise levels for the DGTGV method.

In the CTGV denoising problem (5.11) are two constraint parameters δ_1 and δ_2 given. The first parameter δ_1 is associated with the constrained

$$\|u - u_0\|_2 \leq \delta_1$$

which is the same constraint as in the second step of the DGTV and the DGTGV problems, cf. Problem (4.9).

The second parameter δ_2 belongs to the constraint

$$\||\nabla u - v|\|_1 \leq \delta_2.$$

We can observe the following.

Lemma 5.6

Let

$$u^{TV} \in \arg\min_u \||\nabla u|\|_1 \quad s.t. \quad \|u - u_0\|_2 \leq \delta_1$$

be the solution of the constrained total variation image denoising problem. Then it holds

$$\delta_2 = \||\nabla u_0|\|_1 \Rightarrow u_0 \text{ is a solution of (5.13)},$$
$$\delta_2 = \||\nabla u^{TV}|\|_1 \Rightarrow u^{TV} \text{ is a solution of (5.13)}.$$

Finally, for $\delta_2 = 0$ we obtain that

$$\arg\min_u \{CTGV_0(u) \mid \|u - u_0\|_2 \leq \delta_1\} = \arg\min_u \{\||Hess(u)|\|_1 \mid \|u - u_0\|_2 \leq \delta_1\}.$$

Proof. Let $\delta_2 = \||\nabla u_0|\|_1$. We observe that $(u, v) = (u_0, 0)$ is a feasible and optimal solution of the CTGV problem (5.13) with optimal value zero.

Let $\delta_2 = \||\nabla u^{TV}|\|_1$. Then, we observe that $(u, v) = (u^{TV}, 0)$ is accordingly a feasible and optimal solution, again with optimal value zero.

At last, let $\delta_2 = 0$. Then, $\nabla u = v$ and thus, $\mathcal{E}(v) = \mathcal{E}(\nabla u) = Hess(u)$. $\qquad \square$

What does this mean according to the results of the image denoising? Phrased in other words, if $\delta_2 = \||\nabla u_0|\|_1 = \mathrm{TV}(u_0)$ then the algorithm will not denoise at all. If, on the other hand, $\delta_2 = \||\nabla u^{\mathrm{TV}}|\|_1 = \mathrm{TV}(u^{\mathrm{TV}})$ we would not expect a denoising performance beyond pure total variation denoising. And if $\delta_2 = 0$ the objective function of the problem will become a pure second-order objective. Last, if $\delta_2 > \mathrm{TV}(u^{\mathrm{TV}})$, we get a solution u^{TV} since $(u^{\mathrm{TV}}, 0)$ is still feasible.

Hence, we can only give the heuristic to choose $\delta_2 \in \left[\||\nabla u_0|\|_1, \||\nabla u^{\mathrm{TV}}|\|_1 \right]$. But since we would have to denoise the image first with the pure total variation method to calculate the upper bound of the interval this method seems not to be best according to time performance.

Now, consider the MTGV problem (5.14) with its two problem parameters α and δ_1. The constrained parameter δ_1 is the same as before, we set $\delta_1 = \sigma \|\bar{\eta}\|_2$ where σ is the standard deviation of the noise. Again, we tested various images with different noise levels to determine a range in which the penalization parameter α should be in (see Figure 5.3). It can be seen that the range for α is further spread than in the diagram for the DGTGV penalization parameter (cf. Figure 5.2). In order to get good results for different images and noise levels we set $\alpha = 2$ since $\alpha = 1$ is to small for this problem formulation. Setting $\alpha = 2$ also corresponds with the parameter settings within the TGV problem where $\alpha_0 = 2\alpha_1$.

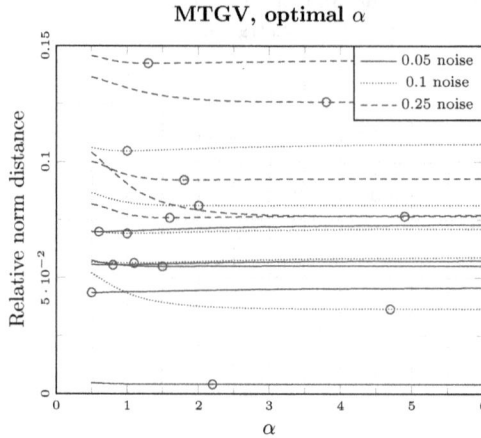

Figure 5.3: Optimal α values for various images and different noise levels for the MTGV method (5.14).

NUMERICS

In the previous chapters we proposed various image denoising methods which can be divided mainly into two groups. The sequential methods and the combined methods similar to the total generalized variation image denoising method (TGV, see [BKP10]). Each of these variational models comes with a set of parameters to choose. In this section we derive numerical methods to solve the minimization problems and try to illustrate the effects of their parameters and evaluate the proposed methods.

We divide the set of parameters into two groups:

Problem parameters: These are the parameters of the model itself. For example in the case of TGV denoising (5.9), the two parameters are the two regularization parameters α_0 and α_1 while for the MTGV denoising (5.14) we have the parameters α and δ_1 and the DGTGV method consisting of (4.10) and (4.9) also has two parameters α and δ_1 (cf. Section 5.5).

Algorithmic parameters: These are parameters that solely influence the algorithm but not the theoretical minimizers. For example, these can be: One or more step sizes, or the stopping criterion (e.g. a tolerance for the duality gap).

6.1 Discretization

In order to implement these various methods we need a few discretized operators for the different derivatives and its adjoints. The discrete gradient $\nabla : \mathbb{R}^{M \times N} \to \mathbb{R}^{M \times N \times 2}$ and the discrete symmetrized gradient $\mathcal{E} : \mathbb{R}^{M \times N \times 2} \to \mathbb{R}^{M \times N \times 3}$ are defined via discrete partial derivatives. As a reminder, they write as follows:

$$\nabla u = \begin{pmatrix} \partial_1 u \\ \partial_2 u \end{pmatrix}, \quad \mathcal{E}(v) = \begin{pmatrix} \partial_1 v_1 & \frac{1}{2}(\partial_1 v_2 + \partial_2 v_1) \\ \frac{1}{2}(\partial_1 v_2 + \partial_2 v_1) & \partial_2 v_2 \end{pmatrix}.$$

For the symmetrized gradient we only save the mixed derivative once in the third layer of the tensor. Let w be this tensor, hence, $w \in \mathbb{R}^{M \times N \times 3}$. For the discrete gradient of $u \in \mathbb{R}^{M \times N}$ we use forward differences, i.e.

$$(\nabla u)_1 = (\partial_1 u), \quad (\nabla u)_2 = (\partial_2 u)$$

with

$$\partial_1, \partial_2 : \mathbb{R}^{M \times N} \to \mathbb{R}^{M \times N},$$

$$(\partial_1 u)_{i,j} = \begin{cases} u_{i+1,j} - u_{i,j}, & \text{if } 1 \le i < M, \\ 0, & \text{if } i = M, \end{cases} \tag{6.1}$$

$$(\partial_2 u)_{i,j} = \begin{cases} u_{i,j+1} - u_{i,j}, & \text{if } 1 \le j < N, \\ 0, & \text{if } j = N. \end{cases}$$

There we have constant boundary extension, $u_{M+1,j} = u_{M,j}$ and $u_{i,N+1} = u_{i,N}$, and for the symmetrized gradient it is

$$(\mathcal{E}(v))_1 = \partial_1 v_1, \quad (\mathcal{E}(v))_2 = \partial_2 v_2, \quad (\mathcal{E}(v))_3 = \frac{1}{2} \left(\partial_2 v_1 + \partial_1 v_2 \right).$$

Further, we will need discrete divergence operators which are (following the notation in [BKP10]) defined as $\text{div} = -\nabla^*$ and $\text{div}^2 = -\mathcal{E}^*$; more detailed we set

$$\text{div}(v) = \partial_1^*(v_1) + \partial_2^*(v_2), \quad \text{div}^2(q) = \begin{pmatrix} \partial_1^*(q_1) + \partial_2^*(q_3) \\ \partial_1^*(q_3) + \partial_2^*(q_2) \end{pmatrix}$$

with

$$(\partial_1^* v)_{i,j} = \begin{cases} -v_{i,j}, & \text{if } i = 1, \\ v_{i,j} - v_{i+1,j}, & \text{if } 1 < i < M, \\ v_{i,j}, & \text{if } i = M - 1, \end{cases}$$

$$(\partial_2^* v)_{i,j} = \begin{cases} -v_{i,j}, & \text{if } j = 1, \\ v_{i,j} - v_{i,j+1}, & \text{if } 1 < j < N, \\ v_{i,j}, & \text{if } j = N - 1. \end{cases}$$

We also need the following norms: $\|\|\cdot\|\|_1$ and $\|\|\cdot\|\|_\infty$ where $|\cdot|$ is the 2-norm. Since we need both norms for $v \in \mathbb{R}^{M \times N \times 2}$ and $w \in \mathbb{R}^{M \times N \times 3}$ we give the discretization of all of them as follows:

$$\|\|v\|\|_1 = \sum_{i=1}^{M} \sum_{j=1}^{N} \sqrt{(v_1)_{i,j}^2 + (v_2)_{i,j}^2},$$

$$\|\|w\|\|_1 = \sum_{i=1}^{M} \sum_{j=1}^{N} \sqrt{(w_1)_{i,j}^2 + (w_2)_{i,j}^2 + 2(w_3)_{i,j}^2},$$

$$\|\|v\|\|_\infty = \max_{i=1,\dots,M} \max_{j=1,\dots,N} \sqrt{(v_1)_{i,j}^2 + (v_2)_{i,j}^2},$$

$$\|\|w\|\|_\infty = \max_{i=1,\dots,M} \max_{j=1,\dots,N} \sqrt{(w_1)_{i,j}^2 + (w_2)_{i,j}^2 + 2(w_3)_{i,j}^2}.$$

Further, we will need several projections according to these norms. Each of the following indicator functions with respect to sets formed from norm balls will be a projection onto the norm ball.

The first projection comes directly from the constraint $\|u - u_0\|_2 \le \delta_1$. This projection we can simply take from Equation (3.14), thus,

$$\text{proj}_{\|\cdot - u_0\|_2 \le \delta_1}(u) = \begin{cases} u, & \text{if } \|u - u_0\|_2 \le \delta_1, \\ u_0 + \delta_1 \frac{u - u_0}{\|u - u_0\|_2}, & \text{else.} \end{cases}$$

The projection onto the $\||\cdot\||_\infty$-ball is discretized as

$$\text{proj}_{\||\cdot\||\leq\alpha}(v) = \frac{v}{\max\left(1, \frac{|v|}{\alpha}\right)},$$

where $|\cdot|$ is again the 2-norm depend on the type of v,

$$|v|_{i,j} = \begin{cases} \sqrt{(v_1)_{i,j}^2 + (v_2)_{i,j}^2}, & \text{if } v \in \mathbb{R}^{M\times N\times 2}, \\ \sqrt{(v_1)_{i,j}^2 + (v_2)_{i,j}^2 + 2(v_3)_{i,j}^2}, & \text{if } v \in \mathbb{R}^{M\times N\times 3}. \end{cases}$$

Then, $|v| \in \mathbb{R}^{M\times N}$.

The last projection that we need is the projection onto the $\||\cdot\||_1$-ball with radius δ_2. Jitkomut Songsiri proposes in [Son11] an efficient method to calculate the Euclidean projection of $\mathbb{R}^n$-vectors onto an ℓ_1-norm ball. The main idea is to solve the dual problem where the optimization variable is simply a scalar. The projection problem writes as

$$\min_y \frac{1}{2} \|y - a\|_2^2 \quad s.t. \quad \|y\|_1 \leq \delta.$$

The Lagrangian is

$$L(y, \lambda) = \|y - a\|_2^2 + 2\lambda(\|y\|_1 - \delta) = \sum_{k=1}^n ((y_k - a_k)^2 + 2\lambda|y_k|) - 2\lambda\delta.$$

To find the dual function they define

$$g_k(\lambda) = \inf_{y_k} (y_k - a_k)^2 + 2\lambda|y_k|$$

where g_k can also be represented by

$$g_k(\lambda) = \begin{cases} -(\lambda - |a_k|)^2 + a_k^2, & \lambda < |a_k|, \\ a_k^2, & \lambda \geq |a_k|. \end{cases}$$

Therefore, the dual function is

$$g(\lambda) = \sum_{k=1}^n g_k(\lambda) - 2\lambda\delta$$

and the dual problem

$$\max_\lambda g(\lambda) \quad s.t. \quad \lambda \geq 0.$$

The optimal value λ^* is given by the root of

$$g'(\lambda) = 2 \sum_{k=1}^n \max(|a_k| - \lambda, 0) - 2\delta$$

which will be calculated numerically. We sort a_k in ascending order, i.e.

$$|a_1| \leq |a_2| \leq \cdots \leq |a_n|,$$

and check in which interval $g'(\lambda)$ changes its sign from a negative to a positive value.

Now, we want to project not onto an ℓ_1-norm ball but rather onto a sum of ℓ_2-norm balls. Here, Songsiri also proposes a method based on the projection onto the ℓ_1-norm ball. Considering $|v|$ instead of v and then projecting this onto the ℓ_1-norm ball does the trick.

We summarize the ideas in Algorithm 2 which follows Algorithm 1 and 2 in [Son11] but using our notation. There, $\tilde{g}' = g'(|a_k|)/2$ and $\tilde{g}'_+ = g'(|a_{k+1}|)/2$.

Algorithm 2: Projection of gradient v onto $\{y \in \mathbb{R}^{m \times n \times 2} \mid |||y|||_1 \le \delta\}$ (cf. [Son11]).

Input: Gradients $v \in \mathbb{R}^{M \times N \times 2}$, $n = M \cdot N$.

Output: Projected gradients y such that $|||y|||_1 \le \delta$.

if $|v| \le \delta$ **then**

$\quad | \quad y = v$

else

$$a = \begin{pmatrix} 0 \\ \text{sort}(\text{vec}(|v|)) \end{pmatrix} = \begin{pmatrix} a_1 \\ \vdots \\ a_{n+1} \end{pmatrix}, \qquad b = \begin{pmatrix} a_{n+1} - \sum_{k=1}^{n} a_k \\ a_{n+1} - \sum_{k=2}^{n} a_k \\ \vdots \\ a_{n+1} - a_n \\ a_{n+1} \end{pmatrix} = \begin{pmatrix} b_1 \\ \vdots \\ b_{n+1} \end{pmatrix}$$

$\quad$ **for** $k = 1, ..., n$ **do**

$$\tilde{g}' = ((k-1) - n)a_k + b_{k+1} - \delta$$
$$\tilde{g}'_+ = \tilde{g}' + ((k-1) - n)|a_{k+1} - a_k|$$

$\qquad$ **if** $\tilde{g}' \ge 0$ *and* $\tilde{g}'_+ \le 0$ **then**

$$\lambda = \frac{\tilde{g}'}{n - (k-1)} + |a_k|$$
$$y = \max\left(1 - \frac{\lambda}{|v|}, 0\right) v$$

$\qquad$ **end**

$\quad$ **end**

$$\lambda = |a_{n+1}| - \delta$$
$$y = \max\left(1 - \frac{\lambda}{|v|}, 0\right) v$$

end

6.2 DGTV and DGTGV

In this section we will give the update functions according to Chambolle-Pock's algorithm (see Algorithm 1 and [CP11]) for the two-stage methods DGTV and DGTGV. Additionally, we will give remarks on the choice of the problem and algorithmic parameters. Later on, we will show results for both methods tested with various images and different noise levels.

Let us start with the DGTV method consisting of the two steps (4.8) and (4.9). Since we want to use the primal-dual Algorithm 1 from Section 2.5 we need to calculate the proximal mappings with respect to the functionals F and G^* and a linear operator K for each of both steps separately. We get the functionals and operators from the problem formulations.

For the first step, Equation (4.8), of the DGTV problem we have

$$F(v) = \mathcal{I}_{\||\nabla u_0 - \cdot\||_1 \le \delta_2}(v),$$
$$G^*(q) = \mathcal{I}_{\||\cdot\||_\infty \le 1}(q),$$
$$K = \mathcal{E}, \quad \text{and} \quad K^* = \mathcal{E}^*.$$

Thus, the proximal mappings are norm ball projections. The updates are

$$
\begin{aligned}
v^{n+1} &= \operatorname{prox}_{\tau F}\left(v^n - \tau K^* q^n\right) \\
&= \operatorname{proj}_{\||\nabla u_0 - \cdot\||\le\delta_2}\left(v^n - \tau\,\mathcal{E}^*\,q^n\right),
\end{aligned}
\tag{6.2}
$$

$$
\begin{aligned}
q^{n+1} &= \operatorname{prox}_{\sigma G^*}\left(q^{n+1} + \sigma K\left(2v^{n+1} - v^n\right)\right) \\
&= \operatorname{proj}_{\||\cdot\||_\infty \le 1}\left(q^n + \sigma\,\mathcal{E}\left(2v^{n+1} - v^n\right)\right).
\end{aligned}
\tag{6.3}
$$

As a stopping criterion we calculate the duality gap and iterate v^n and q^n until a given tolerance value is reached. The duality gap is given by

$$
\begin{aligned}
\operatorname{gap}_{\mathrm{DGTV}}^{(1)}(v, q) = {}&\mathcal{I}_{\||\nabla u_0 - \cdot\||_1 \le \delta_2}(v) + \||\mathcal{E}(v)\||_1 \\
&+ \delta_2\,\||\mathcal{E}^* q\||_\infty - \langle \mathcal{E}^* q, \nabla u_0\rangle + \mathcal{I}_{\||\cdot\||_\infty \le 1}(q).
\end{aligned}
\tag{6.4}
$$

In the DGTGV problem the gradient denoising step (4.10) can be written with functionals F and G^* and a linear operator K, too, here

$$F(v) = \||v - \nabla u_0\||_1,$$
$$G^*(q) = \mathcal{I}_{\||\cdot\||_\infty \le \alpha}(q),$$
$$K = \mathcal{E}, \quad \text{and} \quad K^* = \mathcal{E}^*.$$

We use Moreau's identity (2.28) in order to formulate these updates as projections as well. We need

$$F^*(p) = \mathcal{I}_{\||\cdot\||_\infty \le 1}(p) + \langle p, \nabla u_0\rangle.$$

Thus, the proximal mappings are norm ball projections and the updates are

$$
\begin{aligned}
v^{n+1} &= \operatorname{prox}_{\tau F}\left(v^n - \tau K^* q^n\right) \\
&= v^n + \nabla u_0 - \tau\,\mathcal{E}^*\,q^n - \operatorname{proj}_{\||\cdot\||_\infty \le 1}\left(\tau^{-1}v^n - \mathcal{E}^* q^n\right),
\end{aligned}
\tag{6.5}
$$

$$
\begin{aligned}
q^{n+1} &= \operatorname{prox}_{\sigma G^*}\left(q^{n+1} + \sigma K\left(2v^{n+1} - v^n\right)\right) \\
&= \operatorname{proj}_{\||\cdot\||_\infty \le \alpha}\left(q^n + \sigma\,\mathcal{E}\left(2v^{n+1} - v^n\right)\right).
\end{aligned}
\tag{6.6}
$$

As a stopping criterion we, as above, calculate the duality gap and optimize v^n until a given tolerance value is reached. The duality gap for the DGTGV problem is given by

$$
\begin{aligned}
\mathrm{gap}_{\mathrm{DGTGV}}^{(1)}(v, q) = &\; \||\nabla u_0 - v|\|_1 + \alpha \, \||\mathcal{E}(v)|\|_1 \\
&+ \mathcal{I}_{\||\cdot|\|_\infty \leq 1}(\mathcal{E}^* q) - \langle \mathcal{E}^* q, \nabla u_0 \rangle + \mathcal{I}_{\||\cdot|\|_\infty \leq \alpha}(q).
\end{aligned}
\tag{6.7}
$$

Both methods share the second step (4.9) which we can also write via functionals F and G^* and a linear operator K

$$
\begin{aligned}
F(u) &= \mathcal{I}_{\|\cdot - u_0\|_2 \leq \delta_1}(u), \\
G^*(p) &= \mathcal{I}_{\||\cdot|\|_\infty \leq 1}(p) + \langle p, \hat{v} \rangle, \\
K &= \nabla, \quad \text{and} \quad K^* = \nabla^*,
\end{aligned}
$$

where $\hat{v}$ is the solution of the first step, see (6.2), (6.3) or (6.5), (6.6). Therefore, the updates are

$$
\begin{aligned}
u^{n+1} &= \mathrm{prox}_{\tau F}\left(u^n - \tau K^* p^n\right) \\
&= \mathrm{proj}_{\|\cdot - u_0\|_2 \leq \delta_1}\left(u^n - \tau \nabla^* p^n\right)
\end{aligned}
\tag{6.8}
$$

$$
\begin{aligned}
p^{n+1} &= \mathrm{prox}_{\sigma G^*}\left(p^{n+1} + \sigma K\left(2u^{n+1} - u^n\right)\right) \\
&= \mathrm{proj}_{\||\cdot|\|_\infty \leq 1}\left(p^n + \sigma\left(\nabla\left(2u^{n+1} - u^n\right) - \hat{v}\right)\right).
\end{aligned}
\tag{6.9}
$$

At last, here, we can also use the duality gap as a stopping criterion

$$
\begin{aligned}
\mathrm{gap}_{\mathrm{DGTGV}}^{(2)}(u, p) = &\; \mathcal{I}_{\|\cdot - u_0\|_2 \leq \delta_1}(u) + \||\nabla u - \hat{v}|\|_1 \\
&- \langle \nabla^* p, u_0 \rangle + \mathcal{I}_{\||\cdot|\|_\infty \leq 1}(p) + \langle p, \hat{v} \rangle.
\end{aligned}
\tag{6.10}
$$

We tested both methods with various images and different noise levels. Figure 6.1 shows results of both methods with the default parameters, for DGTV it is $c = 0.99$ and for DGTGV it is $\alpha = 1$ (cf. Section 5.5). The image $u^\dagger$ was corrupted by Gaussian noise of mean 0 and variance 0.1. The corrupted image is u_0.

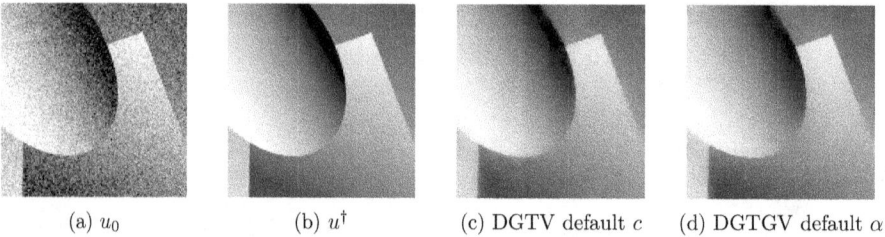

(a) u_0 (b) $u^\dagger$ (c) DGTV default c (d) DGTGV default α

Figure 6.1: Comparison of DGTV and DGTGV methods of the affine image with default c and default α values; noise of mean 0 and variance 0.1.

Both methods with default problem parameters show good reconstruction results. In Figure 6.2 we compare both methods at an skew edge (skew with respect to the x_1 and x_2 coordinates according to the image). Both reconstructions preserve the edge very well.

In Figure 6.3 we look at the same skew edge and compare the gradients before and after denoising with both the DGTV and the DGTGV method. We see that after denoising the gradients within smooth regions, according to the gray value, point roughly in the same direction.

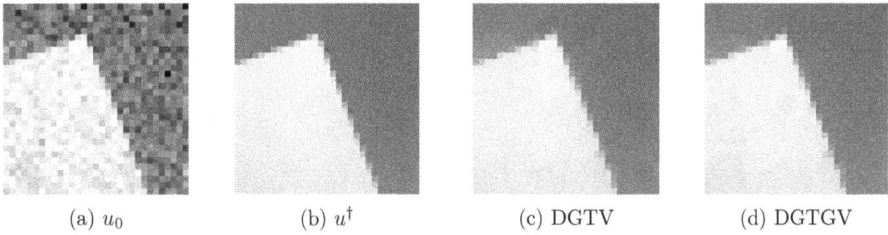

(a) u_0 (b) $u^\dagger$ (c) DGTV   (d) DGTGV

Figure 6.2: Comparison of DGTV and DGTGV methods of the affine image at a tilted edge; noise of mean 0 and variance 0.1.

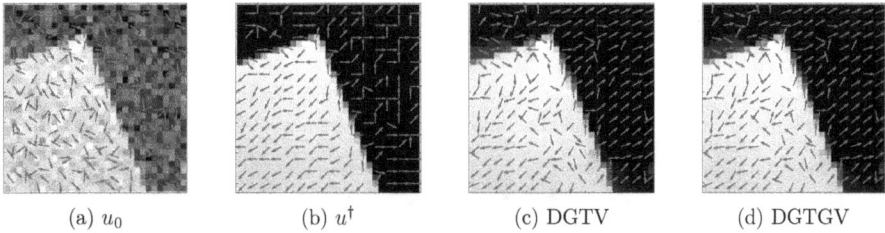

(a) u_0 (b) $u^\dagger$ (c) DGTV   (d) DGTGV

Figure 6.3: Comparison of DGTV and DGTGV methods of the affine image at an edge with gradient vectors before and after computing.

Hence, the denoising processes smooth the gradients and they also point in roughly the same direction as the gradients of the original undisturbed image $u^\dagger$.

In Figure 6.4 we compare the optimization results for the DGTGV method with the default α-value with an optimized problem parameter α. With an optimal α we mean such that the distance of the denoised image to the ground truth noise-free image is as small as possible.

We do the same comparison in Figures 6.5 and 6.6 for different images and noise levels. For the eye image in Figure 6.5 we consider a quite large noise level and in the barbara image (Figure 6.6) a quite small one.

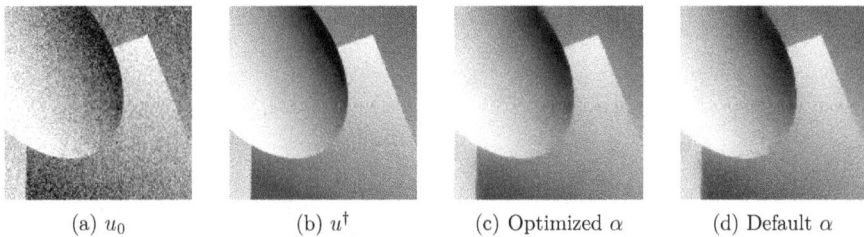

(a) u_0 (b) $u^\dagger$ (c) Optimized α (d) Default α

Figure 6.4: Comparison of DGTGV method of affine image with default α value and best α value; noise of mean 0 and variance 0.1.

There, we see that it is possible to reconstruct the fine stripes that are on the tablecloth or on the trousers if there is a noise level that is small enough.

In all of these images we see that the optical results for DGTGV with default problem parameter α do not differ much from the results with an optimized α value.

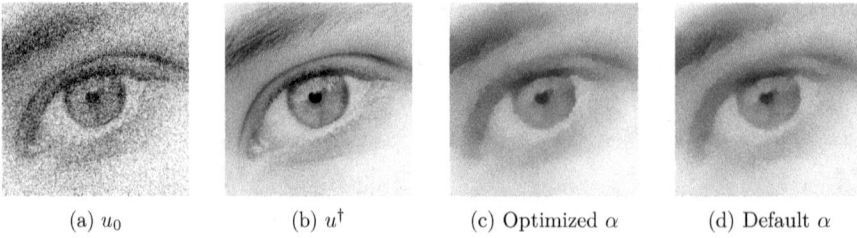

(a) u_0 (b) $u^\dagger$ (c) Optimized α (d) Default α

Figure 6.5: Comparison of DGTGV method of the eye image with default α value and best α value; noise of mean 0 and variance 0.25.

(a) u_0 (b) $u^\dagger$

(c) Optimized α (d) Default α

Figure 6.6: Comparison of DGTGV method of the barbara image with default α value and best α value; noise of mean 0 and variance 0.05.

6.3 CTGV

This section will focus on the CTGV problem, see Equation (5.11). Let us rewrite the CTGV-problem with functionals F, G^* and a linear operator K:

$$F(u,v) = \mathcal{I}_{\|\cdot - u_0\|_2 \leq \delta_1}(u),$$
$$G^*(p,q) = \delta_2 \|\|p\|\|_\infty + \mathcal{I}_{\|\|\cdot\|\|_\infty \leq 1}(q),$$
$$K = \begin{pmatrix} \nabla & -\mathrm{Id} \\ 0 & \mathcal{E} \end{pmatrix}, \quad \text{and} \quad K^* = \begin{pmatrix} \nabla^* & 0 \\ -\mathrm{Id} & \mathcal{E}^* \end{pmatrix}.$$

As a stopping criterion we use the duality gap which is

$$\begin{aligned}
\mathrm{gap}_{\mathrm{CTGV}}(u,v,p,q) = {} & \mathcal{I}_{\|\cdot - u_0\|_2 \leq \delta_1}(u) + \mathcal{I}_{\|\|\cdot\|\|_1 \leq \delta_2}(\nabla u - v) + \|\|\mathcal{E}\,v\|\|_1 \\
& + \delta_1 \|\nabla^* p\|_2 + \langle p, \nabla u_0 \rangle + \mathcal{I}_{\{0\}}(p - \mathcal{E}^* q) \\
& + \delta_2 \|\|p\|\|_\infty + \mathcal{I}_{\|\|\cdot\|\|_\infty \leq 1}(q).
\end{aligned} \tag{6.11}$$

Problematic in this method is the indicator function $\mathcal{I}_{\{0\}}(p - \mathcal{E}^* q)$, since usually the equality $p = \mathcal{E}^* q$ does not hold and thus, the duality gap will not be finite. There are methods to avoid this issue by substituting p and q in the gap or reformulate the problem itself by the introduction of another variable $w = \nabla u - v$.

First, we tested the algorithm under the assumption that $\delta_1 = \|u^\dagger - u_0\|_2$ is known. Since the choice of δ_2 is not easily done we used the heuristic given in Lemma 5.6 and calculated an image u^{TV} as a total variation denoised image. Then we set

$$\delta_2 = \frac{\|\|u^{\mathrm{TV}}\|\|_1}{2}$$

in order to see if the method works.

In Figure 6.7 we see good results for the CTGV method in comparison with the pure total variation denoising method. However, this cannot be said in case of the cameraman image (cf. Figure 6.8). We may argue that this is due to the fact that this image does not contain too many smoothly varying regions but many sharp transitions to that the TV denoising already does a good job (see [KL17]).

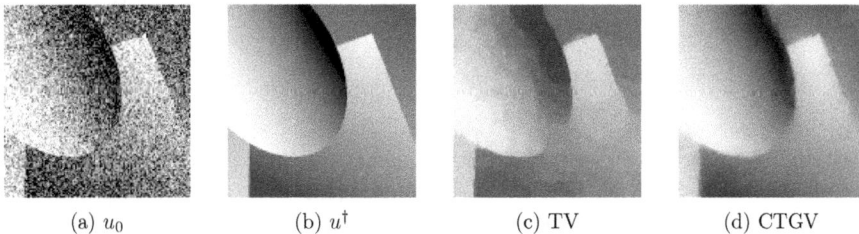

(a) u_0 (b) $u^\dagger$ (c) TV   (d) CTGV

Figure 6.7: Comparison of the total variation denoising and constraint total generalized variation according to the affine image with known δ_1.

Since we did not find a way to easily and reasonably choose δ_2 we will not look further into this method and will go on to the next.

(a) u_0 (b) $u^\dagger$ (c) TV   (d) CTGV

Figure 6.8: Comparison of the total variation denoising and constraint total generalized variation according to the cameraman image with known δ_1.

6.4 MTGV

In this section we evaluate the MTGV method proposed in Section 5.4. We will give the functions used in the algorithm and we will look at the results of the optimization process according
to denoising where we will minimize with default problem parameter α, optimized α values,
and time performance. After that we will compare this method with the previously evaluated
DGTGV method since the MTGV method is in some sense a combination of the two steps of
the DGTGV method. We will compare the performance and the resulting images according to
visual results as well as to PSNR (Peak-Signal-to-Noise-Ratio) values.

Similarly to the previous sections the MTGV problem (5.14) can be written with functionals
F and G^* and a linear operator K such that

$$F(u, v) = \mathcal{I}_{\|\cdot - u_0\|_2 \leq \delta_1}(u),$$

$$G^*(p, q) = \mathcal{I}_{\|\|\cdot\|\|_\infty \leq 1}(p) + \mathcal{I}_{\|\|\cdot\|\|_\infty \leq \alpha}(q),$$

$$K = \begin{pmatrix} \nabla & -\mathrm{Id} \\ 0 & \mathcal{E} \end{pmatrix}, \quad \text{and} \quad K^* = \begin{pmatrix} \nabla^* & 0 \\ -\mathrm{Id} & \mathcal{E}^* \end{pmatrix}.$$

As a stopping criterion we also use the duality gap

$$\mathrm{gap}_{\mathrm{MTGV}}(u, v, p, q) = \|\|\nabla u - v\|\|_1 + \alpha\,\|\|\mathcal{E}(v)\|\|_1 + \mathcal{I}_{\|\cdot - u_0\|_2 \leq \delta_1}(u)$$
$$+\delta_1\,\|\nabla^* p\|_2 - \langle p, \nabla u_0 \rangle + \mathcal{I}_{\{0\}}(p - \mathcal{E}^* q) \qquad (6.12)$$
$$+\mathcal{I}_{\|\|\cdot\|\|_\infty \leq 1}(p) + \mathcal{I}_{\|\|\cdot\|\|_\infty \leq \alpha}(q).$$

The gap is usually not finite as $p = \mathcal{E}^* q$ usually is not fulfilled, but we can circumvent this
in a few steps. First, we replace p by $\mathcal{E}^* q$. The remaining indicator functions do not necessarily
need to be finite, hence, we do a second substitution by replacing p and q in the duality gap
(6.12) by the following terms (see also [KLV18]):

$$\tilde{q} = \frac{q}{\max\left(1, \frac{\|\|\mathcal{E}^* q\|\|_\infty}{\alpha}\right)},$$

$$\tilde{p} = \frac{\mathcal{E}^* q}{\max\left(1, \frac{\|\|\mathcal{E}^* q\|\|_\infty}{\alpha}\right)} = \mathcal{E}^* \tilde{q}.$$

As briefly mentioned in the previous section, we are able to resolve the issue of infinite duality
gaps by a reformulation of the problem itself as we introduce new variables. This holds for a

broad class of problems as well. For the MTGV problem we introduce $w = \nabla u - v$ and replace v in the problem formulation. We obtain

$$\min_{u,w} |||w|||_1 + \alpha \, |||\mathcal{E}(\nabla u - w)|||_1 \quad s.t. \quad \|u - u_0\|_2 \leq \delta_1. \tag{6.13}$$

Here, the functionals and linear operator are

$$F(u, w) = |||w|||_1 + \mathcal{I}_{\|\cdot - u_0\|_2 \leq \delta_1}(u),$$
$$G(\varphi) = \alpha \, |||\varphi|||_1 ,$$
$$K = \mathcal{E} \begin{pmatrix} \nabla & -\mathrm{Id} \end{pmatrix}, \quad \text{and} \quad K^* = \begin{pmatrix} \nabla^* \\ -\mathrm{Id} \end{pmatrix} \mathcal{E}^* .$$

For this problem we get the duality gap

$$\begin{aligned}
\mathrm{gap}_{\mathrm{MTGV}}(u, w, q) = {} & |||w|||_1 + \alpha \, |||\mathcal{E}(\nabla u - w)|||_1 + \mathcal{I}_{\|\cdot - u_0\|_2 \leq \delta_1}(u) \\
& + \delta_1 \|\nabla^*(\mathcal{E}^* q)\|_2 - \langle \mathcal{E}^* q , \nabla u_0 \rangle \\
& + \mathcal{I}_{\||\cdot\||_\infty \leq 1}(\mathcal{E}^* q) + \mathcal{I}_{\||\cdot\||_\infty \leq \alpha}(q).
\end{aligned} \tag{6.14}$$

We ensure the finiteness of the indicator functions by projections into the respective sets, respectively, within the optimization process.

We see that this gap is exactly the gap from above, cf. (6.12), where we did a substitution only in the gap itself instead of in the problem formulation.

This method for the stopping criterion does apply to general problems of the form

$$\min_{u,v} F(u) + G(Av) + H(Bu - v)$$

where A and B are linear (and standard regularity conditions, implying Fenchel-Rockafellar duality is fulfilled). The dual problem is

$$\max_{p,q} -F^*(-B^* p) - H^*(p) - G^*(q) + \mathcal{I}_{\{0\}}(p - A^* q).$$

Replacing p by $A^* q$ leads to

$$\max_p -F^*(-B^* A^* q) - H^*(A^* q) - G^*(q)$$

which is the dual problem of

$$\min_{u,w} F(u) + G(A(Bu - w)) + H(w).$$

Now, let us look at the results. As before we tested the method with various images and different noise levels. Figures 6.9 and 6.10 both show images denoised with the MTGV method. We compare the results based on the proposed default problem parameter $\alpha = 1$ and calculated optimal parameters α. Here, optimal means with respect to the image and noise levels.

We see that the results with respect to the optical impression are pretty similar and that the proposed default value leads to good reconstructions. Later on, we compare the results of MTGV and DGTGV, see Table 6.1. There, one can see the results of both methods according to PSNR-values.

In Figure 6.11 we compare the number of iterations and the runtime of TGV and MTGV. Since the duality gap is not suitable to compare different minimization problems the tolerance is given by

$$\frac{\|u - u^{\mathrm{TGV}}\|_2}{\|u^{\mathrm{TGV}}\|_2}$$

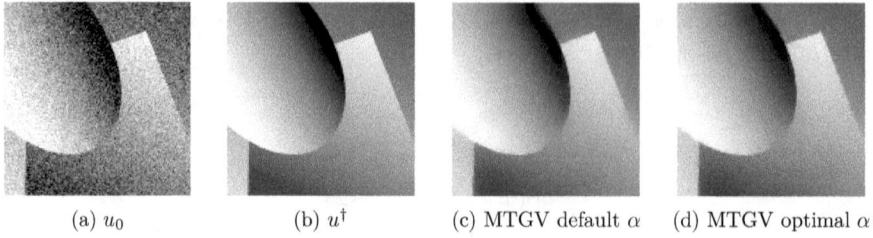

(a) u_0 (b) $u^\dagger$ (c) MTGV default α (d) MTGV optimal α

Figure 6.9: Comparison of MTGV method of the affine image with default α value and best α value; noise of mean 0 and variance 0.1.

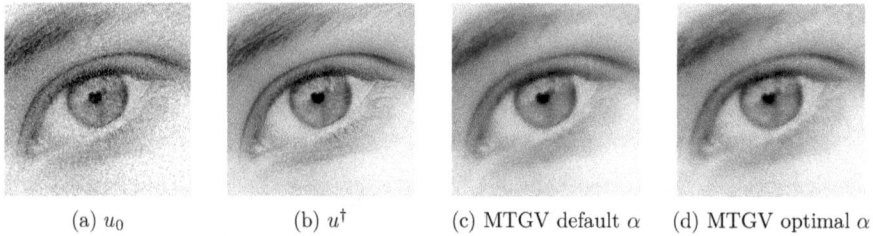

(a) u_0 (b) $u^\dagger$ (c) MTGV default α (d) MTGV optimal α

Figure 6.10: Comparison of MTGV method of the eye image with default α value and best α value; noise of mean 0 and variance 0.1.

where u^{TGV} is the reference value obtained by solving the TGV problem (5.9) with $1\,000\,000$ iterations of Chambolle-Pock (with $\tau = 0.004$, $\sigma = \frac{1}{\tau\|K\|^2}$, $\|K\|^2 = 12$). The algorithms are tested with the eye image (256×256 pixels) corrupted a noise level of 0.1.

We see that the overall behavior of the tolerance given above is pretty similar in both methods. Hence, the methods are competitive.

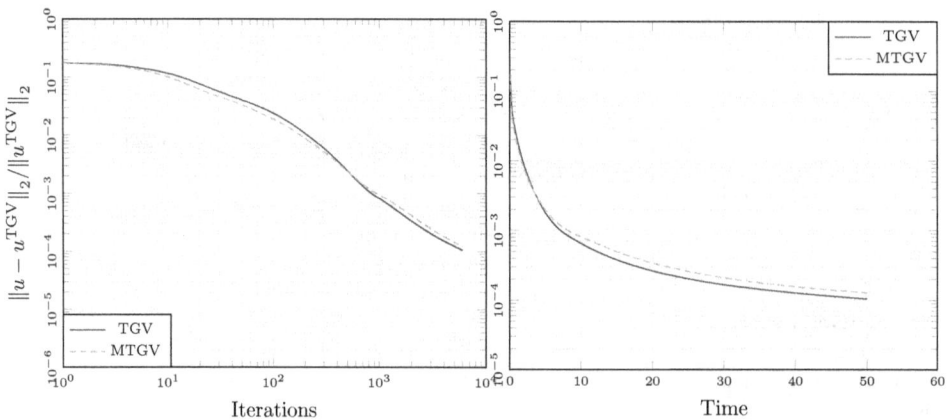

Figure 6.11: Runtime TGV versus MTGV.

6.5 Comparison DGTGV and MTGV

In the previous section we have seen that the DGTGV and the MTGV methods give good image denoising results. As a reminder, both methods use the same functionals

$$\|u - u_0\|_2\,,$$
$$\||\nabla u - v|\|_1\,,$$
$$\text{and} \quad \||\mathcal{E}(v)|\|_1$$

in different ways, i.e. as a penalty or a constraint, respectively. The DGTGV method is a two-stage algorithm that denoises the image gradients first and takes this output gradients as a sort of prior information into the second step of denoising the image itself. The MTGV method does both steps in one and works similarly to the TGV method. A difference is that the discrepancy term $\|u - u_0\|_2$ is not in the minimization functional as a penalty but as a constraint depending on an estimated noise level.

Since both methods have their advantages we compare the performance of both according to quality and runtime.

In Figure 6.12 we used both methods on the affine image which was corrupted by Gaussian noise with mean 0 and standard deviation 0.1.

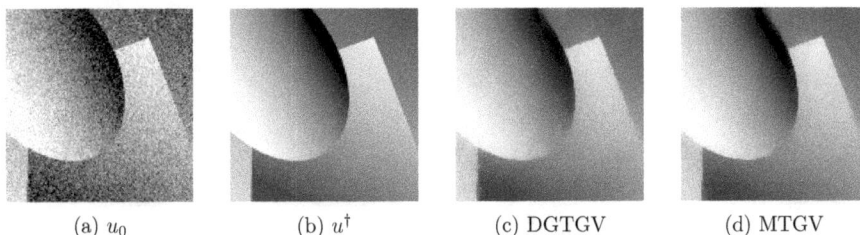

(a) u_0 (b) $u^\dagger$ (c) DGTGV   (d) MTGV

Figure 6.12: Comparison of DGTGV and MTGV method of the affine image with default α value; noise of mean 0 and variance 0.1.

The DGTGV and the MTGV methods both provide good reconstruction results with the respective default problem parameter α. One could argue that in the DGTGV denoised affine image the region where the curvature of the ellipse meets the polygon, the edge is not as sharp as for the MTGV denoised image. But the visual difference is minimal (see Figures 6.13).

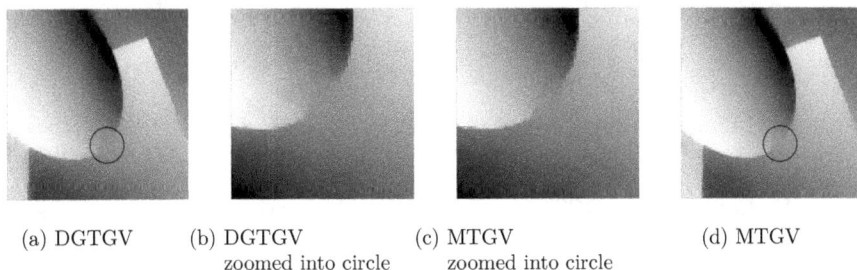

(a) DGTGV   (b) DGTGV   (c) MTGV   (d) MTGV
                   zoomed into circle   zoomed into circle

Figure 6.13: Comparison of DGTGV and MTGV method in one part of the affine image with default α value; noise of mean 0 and variance 0.1.

However, in Figure 6.14 we see, that for both methods the tilted edge is reconstructed very good. All in all we see that the MTGV method leads to a little bit more smooth results than the DGTGV method.

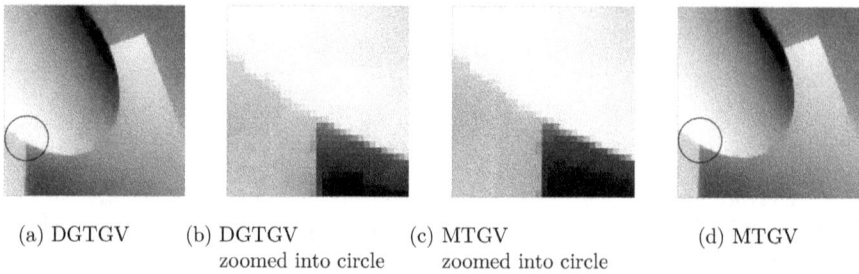

(a) DGTGV   (b) DGTGV   (c) MTGV   (d) MTGV
                 zoomed into circle   zoomed into circle

Figure 6.14: Comparison of DGTGV and MTGV method in one part of the affine image with default α value; noise of mean 0 and variance 0.1.

In Figure 6.15 we consider the same image but now corrupted by noise of a higher level, i.e. 0.25. Even with that much noise both methods provide good denoising results.

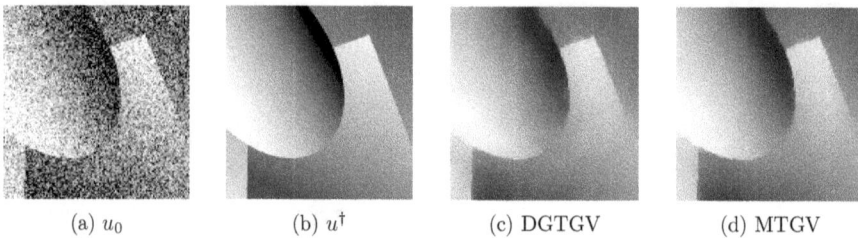

(a) u_0 (b) $u^\dagger$ (c) DGTGV   (d) MTGV

Figure 6.15: Comparison of DGTGV and MTGV method of the affine image with best possible α value; noise of mean 0 and variance 0.25.

For completion we tested and compared both methods with natural images, for example the eye image, cf. Figure 6.16. The difference between the visual results are minimal.

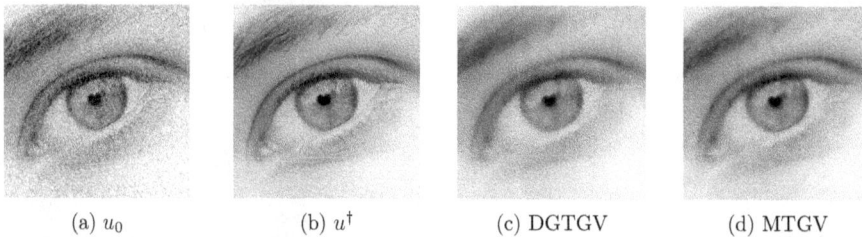

(a) u_0 (b) $u^\dagger$ (c) DGTGV   (d) MTGV

Figure 6.16: Comparison of DGTGV and MTGV method with best possible α value; noise of mean 0 and variance 0.1.

In Table 6.1 we list the PSNR values of images denoised with both methods, DGTGV and MTGV, with both the default and the optimal parameters α with respect to the ground truth.

We see different things in this table. On the one hand, the PSNR values for one method with default values are just slightly smaller than the PSNR values of images denoised with the

Image (variance)	DGTGV	MTGV	DGTGV	MTGV
	best α		default α	
affine (0.05)	37.26	38.30	37.07	38.25
affine (0.1)	32.58	33.97	32.66	33.65
affine (0.25)	27.74	28.41	26.80	27.87
eye (0.05)	31.34	31.64	31.32	31.49
eye (0.1)	28.94	29.37	28.95	29.28
eye (0.25)	26.26	26.98	26.09	26.89
cameraman (0.05)	30.57	30.79	30.53	30.78
cameraman (0.1)	27.14	27.32	27.09	27.32
cameraman (0.25)	23.20	23.34	23.15	23.26
moonsurface (0.05)	30.69	30.87	30.68	30.77
moonsurface (0.1)	28.46	28.73	28.46	28.65
moonsurface (0.25)	26.08	26.38	26.04	26.36
barbara (0.05)	28.35	28.66	28.35	28.49
barbara (0.1)	24.91	25.14	24.91	25.06
barbara (0.25)	22.34	22.48	22.34	22.46

Table 6.1: PSNR values of DGTGV and MTGV methods with best possible α value and default values for each method.

methods with optimal values. On the other hand, we see that the PSNR values according to the DGTGV method and the PSNR values according to the MTGV method also do not differ much.

Thus, both methods provide similar good reconstruction results.

Table 6.2 shows the runtime of both methods. For the DGTGV method we divided the runtime into both steps of the two-stage method in the first two columns. In the third column we list the overall time of the DGTGV method and in the fourth column we list the runtime of the MTGV method. We see that the DGTGV method works slightly faster than the MTGV method. With the barbara image the runtime with the MTGV method is significantly slower than the runtime of the DGTGV method.

All in all, the choice which method one shall use is dependent on what is more important: higher speed or better denoising results. Both methods provide similarly good reconstruction results. When the performance according to runtime is more important than a slightly better PSNR value, we recommend the DGTGV method. If the PSNR value is more important, we recommend the MTGV method.

Image (variance)	DGTGV (grad.)	DGTGV (img.)	DGTGV (all)	MTGV
affine (0.05)	0.12	0.06	0.17	0.20
affine (0.1)	0.15	0.05	0.19	0.25
affine (0.25)	0.27	0.04	0.31	0.39
eye (0.05)	0.14	0.16	0.30	0.91
eye (0.1)	0.23	0.18	0.41	0.93
eye (0.25)	0.67	0.16	0.83	1.53
cameraman (0.05)	0.30	0.15	0.46	1.28
cameraman (0.1)	0.38	0.16	0.54	1.38
cameraman (0.25)	0.73	0.16	0.89	1.81
moonsurface (0.05)	0.14	0.16	0.30	0.84
moonsurface (0.1)	0.28	0.17	0.45	0.93
moonsurface (0.25)	0.65	0.15	0.80	1.52
barbara (0.05)	1.19	0.58	1.77	10.98
barbara (0.1)	1.03	0.60	1.63	7.46
barbara (0.25)	1.94	0.75	2.69	7.33

Table 6.2: Time in seconds for DGTGV and MTGV methods implemented with Chambolle-Pock, for the DGTGV both steps also separately.

SUMMARY

7.1 Summary

This thesis consists of two main parts. The first part is an application of inverse problems for a certain kind of deflectometric measurements. It is based on a collaboration with the Institute of Production Measurement Technology (IPROM) at the Technical University Braunschweig. Deflectometric measurement processes deal with the measurements of specular objects like mirrors or lenses. The datasets provided by the IPROM are sets that describe such objects as a structure. This structure consists of measured surface points and measured surface normals, i.e. the surface's orientation. These two sets, surface points P and normals N, are separate outputs of a deflectometric measurement process [Pet04, Pet06]. Now, the problem at hand is that both sets P and N do not have the same accuracy, which leads to inconsistencies in the dataset. The accuracy of the measured normal vectors is three orders of magnitude higher than that of the point coordinates, since the direct measurement of these points is more sensitive to noise. Thus, we aim at *data fusion*, i.e. to integrate multiple data sources to obtain more accurate data. We propose methods that use the high accuracy of the normal vectors to increase the accuracy of the points in order to have a better description of the measured object.

Instead of using the normals themselves as a kind of "ground truth" information we reformulate these as gradients and solve a minimization problem of the form

$$\min_u \ \||Du - V \nabla u^0|\|_1 \quad s.t. \quad \|u - u^0\|_2 \leq \delta.$$

Here, D represents a discrete differential operator. The matrix V contains directions to neighboring points. The term u^0 represents the third coordinate of the measured points and ∇u^0 are the reformulated normals. We added a constraint $\|u - u^0\|_2 \leq \delta$ where δ gives a tolerance depending on the noise level which can be estimated efficiently. We use the L^1-norm since this norm allows to deal with discontinuities in the measured data and preserves edges.

Our computational results show that the fusion of data greatly improves the accuracy of the point data, which implies that the information of the surface's orientation is quite powerful. We took this insight and used it in image denoising in the second part of this thesis.

Let $u_0 = u^\dagger + \eta$ be an image corrupted by some additive noise η. Considering the total variation denoising problem

$$\min_u \ \lambda \||\nabla u|\|_1 + \frac{1}{2} \|u - u_0\|_2^2$$

we can first reformulate it into a constrained problem like

$$\min_u \ \||\nabla u|\|_1 \quad s.t. \quad \|u - u_0\|_2 \leq \delta_1 \approx \|\eta\|_2$$

and aim to use the information of the surface's orientation as in the case of data fusion. Here, this information would be the "true gradient" $\nabla u^\dagger$ of the uncorrupted image $u^\dagger$. Thus, we formulated the problem

$$\min_{u} \||\nabla u - v|\|_1 \quad s.t. \quad \|u - u_0\|_2 \leq \|\eta\|_2 ,$$

where we want to use gradients v as a prior information. In Lemmas 4.1 and 4.2 we showed that if $v = \nabla u^\dagger$ where known, our method would recover the true image $u^\dagger$. Also, $v = \nabla u^\dagger$ is necessary for this exact recovery. Consequently, we proposed two-stage image denoising methods *Denoised Gradient Total Variation* (DGTV) and *Denoised Gradient Total Generalized Variation* (DGTGV) which calculate an approximation of the true gradient in a first step. The second step takes this approximation and use it as a kind of prior information for the image denoising.

We used functionals that also occur in the total generalized variation (TGV) functional (see [BKP10, KBPS11]), i.e.

$$\text{TGV}_\alpha(u) = \min_{v} \alpha_1 \||\nabla u - v|\|_1 + \alpha_0 \||\mathcal{E}(v)|\|_1 .$$

We combined the functionals

$$\|u - u_0\|_2 ,$$
$$\||\nabla u - v|\|_1 ,$$
$$\text{and} \quad \||\mathcal{E}(v)|\|_1$$

in different ways, i.e. formulated minimization problems where the three functionals occur as penalties or constraints.

Two of the combined problems are a merging of both steps of the two-stage methods. So, combining the steps of the DGTV problem gives us

$$\min_{u,v} \||\mathcal{E}(v)|\|_1 \quad s.t. \quad \|u - u_0\|_2 \leq \delta_1$$
$$\||\nabla u - v|\|_1 \leq \delta_2.$$

We named this problem *Constrained Total Generalized Variation* (CTGV) [KL17]. Combining the steps of the DGTGV problem gives us

$$\min_{u,v} \{ \||\nabla u - v|\|_1 + \alpha \||\mathcal{E}(v)|\|_1 \quad s.t. \quad \|u - u_0\|_2 \leq \delta_1 \}$$

We named this problem *Morozov Total Generalized Variation* (MTGV) [KLV18]. For these combinations we showed that the problems are equivalent to the TGV problem in a certain sense.

All of the combinations come with a set of problem parameters. We discussed the possible choices for these and gave a recommendation as to which problems to consider further, because for some problems easy parameter choice rules exist.

Then, we evaluate the problems in experiments. First, we denoised images with the DGTV, DGTGV, CTGV and the MTGV methods with chosen default problem parameters. We compared the resulting images with those we denoised with specially calculated optimal parameters. The accuracy of the results was good in both cases and close to each other according to PSNR values.

After that we compared the DGTGV and MTGV methods. Those problem formulations consists of the same functionals that are used in the same way according to penalties and constraints. The difference is that DGTGV is a two-stage method and MTGV solves the problem in one step. We saw that the resulting images are visually not very different. This is also confirmed by the PSNR values. The MTGV method provides images with slightly better PSNR values than the DGTGV method. The advantage of the DGTGV method is the runtime. Because of the partition of the problem into two parts the method works faster than the MTGV method.

7.2 Zusammenfassung

Diese Arbeit besteht aus zwei größeren Teilen. Im ersten Teil untersuchen wir eine Anwendung inverser Probleme für eine bestimmte Art deflektometrischer Messungen. Dies basiert auf einer Zusammenarbeit mit dem Institut für Produktionsmesstechnik (IPROM) an der Technischen Universität Braunschweig. Bei deflektometrischen Messprozessen geht es darum, spiegelnde Objekte wie etwa Linsen oder kleine Spiegel auszumessen. Die Daten, die das IPROM bereitgestellt hat, sind Mengen, die gemessene Objekte als Struktur beinhalten. In diesen Strukturen sind gemessene Oberflächenpunkte und gemessene Oberflächennormalen enthalten. Diese Normalenvektoren geben die Orientierung des Objekts im Raum wieder. Diese beiden Mengen, Objektpunkte P und Objektnormalen N, sind separate Ergebnisse aus einem deflektometrischen Prozess [Pet04, Pet06]. Ein Problem ist, dass die Genauigkeit der gemessenen Datenmengen P und N nicht übereinstimmt, was zu Inkonsistenzen innerhalb der Daten führt. Die Genauigkeit der Normalen liegt drei Größenordnungen über derjenigen der Objektpunkte. Dies liegt daran, dass die direkte Messung der Punkte anfälliger für Rauschen ist. Daher ist das Ziel die Daten durch einen Optimierungsprozess so zu kombinieren, dass ihre Genauigkeit erhöht wird. Wir entwickeln Methoden, welche die hohe Genauigkeit der Objektnormalen nutzen, um die Genauigkeit der Punkte zu erhöhen. Dadurch werden die ausgemessenen Objekte besser durch die Daten beschrieben.

Statt die Normalen selbst als Referenzwert zu benutzen, formulieren wir diese als Gradienten um und lösen ein Minimierungsproblem der Form

$$\min_{u} \; \||Du - V\,\nabla\,u^0|\|_1 \quad s.t. \quad \|u - u^0\|_2 \le \delta.$$

Dabei ist D ein diskreter Differentialoperator. Die Matrix V enthält die Richtungen zu benachbarten Punkten. Der Term u^0 ist die dritte Koordinate der gemessenen Punkte und ∇u^0 sind die zu Gradienten umgeschriebenen Normalen. Wir fügen eine Nebenbedingung der Form $\|u - u^0\|_2 \le \delta$ hinzu, wobei δ ein Toleranzbereich der Abweichung zu den gemessenen Daten ist. Der Toleranzbereich hängt dabei vom Rauschen ab. Weiter nutzen wir die L^1-Norm, da diese Unstetigkeiten innerhalb der gemessenen Daten besser handhabt und Kanten erhält.

Unsere Ergebnisse zeigen, dass die Zusammenführung der Daten die Genauigkeit der Objektpunkte stark verbessert. Dies zeigt auch, dass die Orientierung des Objekts eine mächtige Information darstellt. Diese Einsicht nutzen wir im zweiten Teil der Arbeit für das Entrauschen von Bildern.

Sei $u_0 = u^\dagger + \eta$ ein Bild, welches durch additives Rauschen η gestört ist. Betrachten wir nun das Entrauschungsproblem mittels totaler Variation

$$\min_{u} \; \lambda\,\||\nabla\,u|\|_1 + \frac{1}{2}\,\|u - u_0\|_2^2.$$

Dieses können wir in ein Problem mit Nebenbedingung umformulieren,

$$\min_u \, \||\nabla\,u|\|_1 \quad s.t. \quad \|u - u_0\|_2 \leq \delta_1 \approx \|\eta\|_2\,,$$

und wollen hier nun die Information über die Oberflächenorientierung integrieren, wie bei der Datenzusammenführung bei deflektometrischen Messungen. In diesem Fall ist diese Information repräsentiert durch den wahren Gradienten $\nabla\,u^\dagger$ des rauschfreien Bildes $u^\dagger$. Wir ergänzen das Problem also zu

$$\min_u \, \||\nabla\,u - v|\|_1 \quad s.t. \quad \|u - u_0\|_2 \leq \|\eta\|_2\,,$$

wobei wir Gradienten v als eine Art Vorinformationen benutzen möchten. In Lemma 4.1 und Lemma 4.2 zeigen wir, dass falls $v = \nabla\,u^\dagger$ wäre, unsere Methode das rauschfreie Bild $u^\dagger$ wiederherstellen würde. Weiter ist $v = \nabla\,u^\dagger$ auch eine notwendige Bedingung für eine exakte Rekonstruktion. Folglich stellen wir zwei Entrauschungsmethoden vor, *Denoised Gradient Total Variation* (DGTV) und *Denoised Gradient Total Generalized Variation* (DGTGV). Beide Methoden sind Zwei-Schritt-Verfahren. Im ersten Schritt werden die Gradienten des rauschfreien Bildes approximiert und diese dann im zweiten Schritt als eine Art von Vorinformation für das eigentliche Bildentrauschen benutzt.

Weiter, nutzen wir Funktionale, die ebenfalls im Funktional der verallgemeinerten Totalvariation (TGV) vorkommen (siehe [BKP10, KBPS11]), welches

$$\mathrm{TGV}_\alpha(u) = \min_v \, \alpha_1 \, \||\nabla\,u - v|\|_1 + \alpha_0 \, \||\mathcal{E}(v)|\|_1$$

ist. Wir kombinieren die vorkommenden Funktionale

$$\|u - u_0\|_2\,,$$
$$\||\nabla\,u - v|\|_1\,,$$
$$\text{und} \quad \||\mathcal{E}(v)|\|_1$$

auf verschiedene Art, sprich, wir benutzen sie entweder als Nebenbedingung oder Strafterm innerhalb der Minimierungsprobleme.

Zwei der möglichen kombinierten Probleme entstehen direkt, wenn wir die Schritte der Zwei-Schritt-Verfahren zu einem Schritt zusammenfügen. Damit entsteht durch Kombination der Schritte von DGTV das Problem

$$\min_{u,v} \, \||\mathcal{E}(v)|\|_1 \quad s.t. \quad \|u - u_0\|_2 \leq \delta_1$$
$$\||\nabla\,u - v|\|_1 \leq \delta_2.$$

Dieses Problem taufen wir *Constrained Total Generalized Variation* (CTGV) [KL17].

Ähnlich erhalten wir durch Kombination der Schritte von DGTGV das Problem

$$\min_{u,v} \, \{\||\nabla\,u - v|\|_1 + \alpha \, \||\mathcal{E}(v)|\|_1 \quad s.t. \quad \|u - u_0\|_2 \leq \delta_1\}$$

Dieses nennen wir *Morozov Total Generalized Variation* (MTGV) [KLV18]. Für beide Kombinationsprobleme zeigen wir, dass diese auf gewisse Art äquivalent zum TGV-Problem sind.

Letztlich untersuchen wir die vorgestellten Probleme experimentell. Zuerst entrauschen wir Bilder mit den Methoden DGTV, DGTGV, CTGV und MTGV. Hierbei benutzen wir hergeleitete Standardwerte für die Problemparameter. Die Ergebnisse für jede Methode vergleichen wir mit optimal berechneten Problemparametern, die vom Bild abhängen. Die Genauigkeit der Ergebnisse ist für beide Fälle für alle Methoden gut und nah beieinander bezüglich der PSNR-Werte. Danach vergleichen wir die Methoden DGTGV und MTGV. Beide Probleme benutzen baugleiche Funktionale auf die gleiche Weise, d.h. als Nebenbedingung oder Strafterm gleiche Weise. Die entrauschten Bilder haben optisch kaum Unterschiede. Dies wird auch durch die PSNR-Werte bestätigt. Die MTGV Methode liefert leicht bessere Werte als die DGTGV Methode. Der Vorteil der DGTGV Methode ist die schnellere Laufzeit, die durch die Aufteilung in Teilprobleme zu erklären ist.

Bibliography

[AD04] Gabriel Acosta and Ricardo G Durán. An optimal Poincaré inequality in L^1 for convex domains. *Proceedings of the American Mathematical Society*, 132(1):195–202, 2004.

[AFP00] Luigi Ambrosio, Nicola Fusco, and Diego Pallara. *Functions of boounded varation and free discontinuity problems*. The Clarendon Press Oxford University Press, New York, 2000.

[BB03] Jean Bourgain and Haïm Brezis. On the equation div$Y = f$ and application to control of phases. *Journal of the American Mathematical Society*, 16(2):393–426, 2003.

[BG12] Richard L. Bishop and Samuel I. Goldberg. *Tensor Analysis on Manifolds*. Courier Corporation, 2012.

[BKP10] Kristian Bredies, Karl Kunisch, and Thomas Pock. Total generalized variation. *SIAM J. Img. Sci.*, 3(3):492–526, September 2010.

[Bre] Haïm Brezis. *Functional Analysis, Sobolev Spaces and Partial Differential Equations*.

[BV04] Stephen Boyd and Lieven Vandenberghe. *Convex Optimization*. Cambridge University Press, New York, NY, USA, 2004.

[CNCP10] Antonin Chambolle, Matteo Novaga, Daniel Cremers, and Thomas Pock. An introduction to Total Variation for Image Analysis. In *in Theoretical Foundations and Numerical Methods for Sparse Recovery, De Gruyter*, 2010.

[CP11] Antonin Chambolle and Thomas Pock. A first-order primal-dual algorithm for convex problems with applications to imaging. *J. Math. Imaging Vis.*, 40(1):120–145, May 2011.

[ET76] Ivar Ekeland and Roger Temam. *Convex analysis and variational problems*. SIAM, 1976.

[Gau59] Walter Gautschi. Some Elementary Inequalities Relating to the Gamma and Incomplete Gamma Function. *Journal of Mathematics and Physics*, 38(1-4):77–81, 1959.

[GG84] Stuart Geman and Donald Geman. Stochastic relaxation, gibbs distributions, and the bayesian restoration of images. *IEEE Transactions on Pattern Analysis and Machine Intelligence*, PAMI-6(6):721–741, Nov 1984.

[Had02] Jacques Hadamard. Sur les problèmes aux dérivés partielles et leur signification physique. *Princeton University Bulletin*, 13:49–52, 1902.

[Iva62]   Valentin Konstantinovich Ivanov. On linear problems which are not well-posed. *Dokl. Akad. Nauk SSSR*, 145:270–272, 1962.

[KBPS11] Florian Knoll, Kristian Bredies, Thomas Pock, and Rudolf Stollberger. Second order total generalized variation (tgv) for mri. *Magnetic Resonance in Medicine*, 65(2):480–491, 2011.

[KL17]   Birgit Komander and Dirk A. Lorenz. Denoising of image gradients and constrained total generalized variation. In Françoiseditors, *Scale Space and Variational Methods in Computer Vision: 6th International Conference, SSVM 2017, Kolding, Denmark, June 4-8, 2017, Proceedings*, pages 435–446, Cham, 2017. Springer International Publishing.

[KLF+14] Birgit Komander, Dirk Lorenz, Marc Fischer, Marcus Petz, and Rainer Tutsch. Data fusion of surface normals and point coordinates for deflectometric measurements. *J. Sens. Sens. Syst.*, pages 281–290, 2014.

[KLV18]   Birgit Komander, Dirk A. Lorenz, and Lena Vestweber. Denoising of image gradients and total generalized variation denoising. *Journal of Mathematical Imaging and Vision*, May 2018.

[Kom13]   Birgit Komander. Ausgleichsprobleme in der Photogrammetrie. Master's thesis, TU Braunschweig, 2013.

[LOT04]   Marius Lysaker, Stanley Osher, and Xue-Cheng Tai. Noise removal using smoothed normals and surface fitting. *IEEE Trans. Img. Proc.*, 13(10):1345–1357, October 2004.

[LRKB13] Thomas Luhmann, Stuart Robson, Stephen Kyle, and Jan Boehm. *Close-range Photogrammetry and 3D Imaging*. De Gruyter textbook. De Gruyter, 2013.

[LTO12]   Xinhao Liu, Masayuki Tanaka, and Masatoshi Okutomi. Noise level estimation using weak textured patches of a single noisy image. In *Image Processing (ICIP), 2012 19th IEEE International Conference on*, pages 665–668. IEEE, 2012.

[LTO13]   Xinhao Liu, Masayuki Tanaka, and Masatoshi Okutomi. Single-image noise level estimation for blind denoising. *IEEE transactions on image processing*, 22(12):5226–5237, 2013.

[Luh00]   Thomas Luhmann. *Nahbereichsphotogrammetrie: Grundlagen, Methoden und Anwendungen*. Wichmann, 2000.

[LW13]   Dirk A. Lorenz and Nadja Worliczek. Necessary conditions for variational regularization schemes. *Inverse Problems*, 29:075016pp, 2013.

[Min62]   George J. Minty. Monotone (nonlinear) operators in a Hilbert space. *Duke Mathematical Journal*, 29:341–346, 1962.

[Mor67]   Vladimir Alekseevich Morozov. Choice of parameter in solving functional equations by the method of regularization. *Dokl. Akad. Nauk SSSR*, 175:1225–1228, 1967.

[MS89]   David Mumford and Jayant Shah. Optimal approximations by piecewise smooth functions and associated variational problems. *Communications on Pure and Applied Mathematics*, 42(5):577–685, July 1989.

[Pet04] Marcus Petz. Rasterreflexions-Photogrammetrie zur Messung spiegelnder Ober-
         flächen. *tm - Technisches Messen*, 71:389–397, 2004.

[Pet06] Marcus Petz. *Rasterreflexions-Photogrammetrie - Ein neues Verfahren zu ge-
         ometrischen Messung spiegelnder Oberflächen.* Dissertation, Technische Universität
         Braunschweig, 2006.

[Roc70] R. Tyrrell Rockafellar. *Convex analysis.* Princeton Mathematical Series. Princeton
         University Press, Princeton, N. J., 1970.

[ROF92] Leonid I. Rudin, Stanley Osher, and Emad Fatemi. Nonlinear total variation based
         noise removal algorithms. *Physica D.*, 60(1-4):259–268, November 1992.

[Rud91] Walter Rudin. *Functional Analysis.* International series in pure and applied math-
         ematics. McGraw-Hill, 1991.

[Son11] Jitkomut Songsiri. Projection onto an l_1-norm ball with application to identification
         of sparse autoregressive models. *Asean Symposium on Automatic Control*, 2011.
         Vietnam.

[TA77]   Andrei N. Tikhonov and Vasiliy Y. Arsenin. *Solutions of ill-posed problems.* Scripta
         series in mathematics. Winston, 1977.

[Tik63] Andrei N. Tikhonov. Solution of incorrectly formulated problems and the regular-
         ization method. *Soviet Math. Dokl.*, 4:1035–1038, 1963.

$\mathrm{BGV}_\alpha^k(\Omega)$	Space of bounded total generalized variation, page 21
$\mathrm{BV}(\Omega)$	Space of functions with bounded total variation, page 19
$\mathcal{N}(i,j)$	Neighborhood of a pixel $p_{i,j}$, page 37
$\overline{B}_{X,r}(x)$	Closed ball in X around x with radius r, page 8
CTGV	Constrained total generalized variation, page 69
DGTGV	Denoised Gradient Total Generalized Variation, page 65
DGTV	Denoised Gradient Total Variation, page 65
dom F	Domain of a functional F, page 12
∇u^0	Gradient vectors of measured surface u^0 as reformulation of normals, page 39
graph F	Graph of a functional F, page 12
$\mathcal{I}_C(x)$	Indicator function of a set C, page 13
ker F	Kernel of a functional F, page 12
MTGV	Morozov total generalized variation, page 70
$\|\cdot\|_{L^p(\Omega)}$	Norm in $L^p(\Omega)$, page 6
$\|K\|_{X\to Y}$	Operator norm of K, page 6
$\|\cdot\|_X$	Norm on vector space X, page 5
$\boldsymbol{N}$	Tensor of measured normal vectors as of type $M \times N \times 3$, page 52
∂F	Subdifferential of F, page 14
$\boldsymbol{P}$	Tensor of measured point-coordinates as of type $M \times N \times 3$, page 52
prox_F	Proximal operator with respect to F, page 28
ran F	Range of a functional F, page 12
$\langle\,\cdot\,,\,\cdot\,\rangle_X$	Inner product on X, page 9
$\langle x^*,\,x\rangle_{X^* \times X}$	Dual pair, page 7
$\langle\,\cdot\,,\,\cdot\,\rangle$ $\langle\,\cdot\,,\,\cdot\,\rangle_{L^2(\Omega)}$	Standard inner product on $L^2(\Omega)$, page 9

$\|u\|_{W^{m,p}(\Omega)}$ Norm in Sobolev spaces, page 11

$\mathcal{E}(v)$ Symmetrized Jacobian of v, page 65

TGV_α^k Total generalized variation functional of order k and with weights α, page 21

$W^{m,p}(\Omega)$ Sobolev space of order m, p, page 10

A^+ Pseudo-inverse of A, page 41

$B_{X,r}(x)$ Open ball in X around x with radius r, page 8

$D_h F(x)$ Directional derivative of F at x in direction h, page 9

F^* convex conjugate of F, page 16

$H^m(\Omega)$ Sobolev space $W^{m,2}(\Omega)$, page 11

$I_{i,j}$ Set of indices for neighboring points of $p_{i,j}$, page 37

$L^2(\Omega)$ Hilbert space of square-integrable functions on Ω, page 9

$L^p(\Omega)$ Banach space of p-integrable functions on Ω, page 6

$L^\infty(\Omega)$ Banach space of essentially bounded measurable functions on Ω, page 6

$v_{k,\ell}$ Directional vectors point from $p_{i,j}$ to its neighbors, page 41

$x_n^* \overset{*}{\rightharpoonup} x^*$ Weak-* convergence for $x_n^*, x^* \in X^*$, page 8

$x_n \rightharpoonup x$ Weak convergence for $x_n, x \in X$, page 8

www.ingramcontent.com/pod-product-compliance
Lightning Source LLC
Chambersburg PA
CBHW081110220326
41598CB00038B/7295